The Science of *Aruṇapraṣna:*

सूर्यो॑ नो दि॒वस्पा॑तु॒ वातो॑ अं॒तरि॑क्षात् ।
अ॒ग्निर्नः॒ पार्थि॑वेभ्यः ॥

जोषा॑ सवित॒र्यस्य॑ ते॒ हरः॑ श॒तं स॒वाँ अर्ह॑ति ।
पा॒हि नो॑ दि॒द्युतः॒ पतं॑त्याः ॥

चक्षु॑र्नो दे॒वः स॑वि॒ता चक्षु॑र्न उ॒त पर्व॑तः ।
चक्षु॑र्धा॒ता द॑धातु नः ॥

चक्षु॑र्नो धेहि॒ चक्षु॑षे॒ चक्षु॑र्वि॒ख्यै त॒नूभ्यः॑ ।
सं चे॒दं वि च॑ पश्येम ॥

सु॒सं॒दृशं॑ त्वा व॒यं प्रति॑ पश्येम सूर्य ।
वि प॑श्येम नृ॒चक्ष॑सः ॥

ऋगवेदः

१० मण्डलम् १५८ सूक्ता

८ अष्टका ८ अध्यायः १६ वर्गः

१० मण्डलम् १२ अनुवाकः

The Science of *Aruṇapras̰na:*

Cosmology to Quantum Physics

First Edition

Sitaram Ayyagari

Kovidanam Vani - A Rays LLC Company

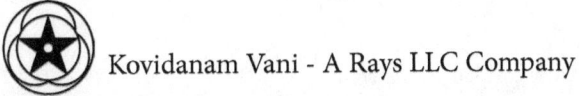

Published in 2022 by Kovidanam Vani, Wilmington 19807

ISBN 978-0-9817596-0-9 (Paperback)

ISBN 978-0-9817596-4-7 (ePub)

Kovidanam Vani acknowledges Mr. Krupalu Vogeti, Ms. Gayatri Nagamani Velpucherla, and Mr. M. V. R. Amarendra Nath for the Sanskrit translation of the publisher's note and Ms. Pavani Sairam Uppuluri for typed Sanskrit text of Aruna Prasna in Devanagari Script. Author acknowledges Srikari Ayyagari for her valuable support and help in publishing this book and Mr. Prasad Ponnapalli for cover page illustration and for the illustration on page 124. Images on pages 10, 12, 14, 16, 38, 44, 50, 54, 56, 61, 68, 74, 79, 87, 89, 92, 96, 112, and 116 are not copyrighted by Kovidanam Vani. They are reproduced in this book under the public domain or various types of creative commons licenses as notated in the caption below the images or text on these pages, respectively.

Publisher's Cataloging-in-Publication (Provided by Cassidy Cataloguing Services, Inc.).

Names: Ayyagari, Sitaram, author.

Title: The science of Aruṇapraṣna : cosmology to quantum physics / Sitaram Ayyagari.

Description: First edition. | Wilmington, [Delaware] : Kovidanam Vani, a Rays LLC company, [2022] | Includes bibliographical references and index.

Identifiers: ISBN: 978-0-9817596-0-9 (paperback) | 978-0-9817596-4-7 (ePub) | LCCN: 2022940301

Subjects: LCSH: Vedas and science. | Aranyakas. Taittirīyāraṇyaka. Aruṇapraśna--Criticism, interpretation, etc. | Vedas--Criticism, interpretation, etc. | Science--Religious aspects--Hinduism. | Cosmology--Religious aspects--Hinduism. | Quantum theory--Religious aspects--Hinduism. | Atmosphere. | Rain and rainfall. | Gravity. | Solar radiation.

Classification: LCC: BL1122.27 .A99 2022 | DDC: 294.5921--dc23

www.kovidanamvani.com

To my parents

Sri Sreerama Murty & Smt. Nagaratnam Ayyagari

Except for Publisher's Note and the Sanskrit Transliteration, all information including hypotheses, analyses and opinions presented in this book are of the author, Mr. Sitaram Ayyagari's only and not of Kovidanam Vani or Rays LLC or anyone else.

Contents

Kovidanam Vani Alphabet of Sanskrit Transliteration १४

प्रकाशकस्य टिप्पणी १९

Forewords २३

Introduction 1

Introduction 3

Ḇūmī 9

Science 11

Aruṇapraṣna 19

Ḇūmī Postulates 31

Ḇūmī Postulate 2.1 33

Ḇūmī Postulate 2.2 35

Ḇūmī Postulate 2.3 39

Conclusion 41

Vāyumaṇḍalam 43

Science 45

Aruṇapraṣna 47

Vāyumaṇḍalam Postulates 49

Vāyumaṇḍalam Postulate 3.1 51

Vāyumaṇḍalam Postulate 3.2 55

Vāyumaṇḍalam Postulate 3.3 57

Conclusion 58

Varṣā 59

Science 60

Aruṇapraṣna 63

Varṣā Postulates 67

Varṣā Postulate 4.1 69

Varṣā Postulate 4.2 72

Varṣā Postulate 4.3 75

Conclusion 76

Gurutvākarṣaṇa 77

Science 78

Aruṇapraṣna 82

Gurutvākarṣaṇa Postulates 86

Gurutvākarṣaṇa Postulate 5.1 88

Gurutvākarṣaṇa Postulate 5.2 90

Gurutvākarṣaṇa Postulate 5.3 93

Conclusion 94

Sūryaraṣmi 95

Science 97

Aruṇapraṣna 99

Sūryaraṣmi Postulates 107

Sūryaraṣmi Postulate 6.1 109

Sūryaraṣmi Postulate 6.2 111

Sūryaraṣmi Postulate 6.3 113

Conclusion 114

Yajṇenabanḍu 115

Science 117

Yajṇenabanḍu Postulates 123

Yajṇenabanḍu Postulate 7.1 125

Yajṇenabanḍu Postulate 7.2 138

Yajṋenabanḍu Postulate 7.3 143

Conclusion 144

Glossary of Selected Sanskrit Terms 147

Glossary of Selected Scientific Terms 172

Text of *Aruṇaprasṉa:* 182

Bibliography 213

Index 219

About Author 242

Illustrations

A representation of the evolution of the universe over 13.77 billion years. 10

An artist's imagination of cyclical Big Bounce type universe. 10

Artist's concept of a protoplanetary disk, where particles of dust and grit collide and accrete forming planets or asteroids. 12

Backjet of a drop of water after impact on a water-surface. 14

Timeline of the evolutionary history of life. 16

Hindu Cosmology is based on infinite cycles of time. 18

Aruṇapraṣna's Cosmology of the Earth. *Āpa* emerges from *Salilam* and the physcial Earth comes out of *Āpa*. 32

Aruṇapraṣna's Cosmology of the Earth from *Salilam* to Lifeforms. 34

Earth's axis of rotation changes from North - South to the present slanted Northeast -Southwest. 38

Layers of the atmosphere. 44

Diagram illustrating the ozone-oxygen cycle. 50

Pravargya Ritual in Peddapuram, AP, Bharat. 54

Proposed Atmospheric Layers based on *Veda* mapped to the Atmospheric Layers based on Science. 56

USGS, Graph of the locations of water on Earth. 61

Isolated Towering Vertical Thunderhead in the Mojave Desert. 68

Vāyugaṇas, Condensation Quantum Fields. 71

Aerial view of the Amazon Rainforest, near Manaus, the capital of the Brazilian state of Amazonas. 74

Lattice analogy of the deformation of spacetime caused by a planetary mass. 79

Viṣṇu in *Varāha Avatāra*. 87

Artist's imagination of Gravity Fields, Space-Time Fabrics or Quantum Gravity Fields as *Vāyu* and *Agni*. 89

Water-Cycle Diagram. 92

Image showing the distribution of electromagnetic waves with respect to frequency and wavelength, highlighting visible part of the electromagnetic spectrum. 96

Saptasūryas, EMR Spectrum. 108

Soma enhances *Saptasūryas*. 110

Aruṇapraṣna's EMR labels superimposed on an existing NASA image. 112

Cush, Standard model of elementary particles: the 12 fundamental fermions and 5 fundamental bosons. 116

Yajṇa is a Quantum Entanglement Engine. 124

Approximate layout of a *Ṣrota Vihāra*. Not to scale. Please refer to the legend provided at the end of these series of images. 129

Approximate layout of the *Pracīnāva'ṣa* section (Rectangle labeled number 4) of the *Ṣrota Vihāra*. Not to scale. Please refer to the legend provided at the end of these series of images. 130

Approximate layout of the *Mahāvedi* section (Trapezium labeled number 1) of the *Ṣrota Vihāra*. Not to scale. Please refer to the legend provided at the end of these series of images. 131

Approximate layout of the *Uttaravedi Citti* section (Rectangle labeled number 11) of the *Mahāvedi*. Not to scale. Please refer to the legend provided at the end of these series of images. 132

Approximate cross-sectional layout of the *Uttaravedi Citti* section (Rectangle labeled number 11) of the *Mahāvedi*. Not to scale. Please refer to the legend provided at the end of these series of images. 133

Kovidanam Vani Alphabet of Sanskrit Transliteration

D̉evanāgarī	**Roman Lower Case**	**Roman Upper Case**
ॐ	ॐ	ॐ
अ	*a*	*A*
आ	*ā*	*Ā*
इ	*i*	*I*
ई	*ī*	*Ī*
उ	*u*	*U*
ऊ	*ū*	*Ū*
ऋ	*r̰*	*R̰*
ॠ	*r̰̄*	*R̰̄*
ऌ	*l̰*	*L̰*
ए	*e*	*E*
ऐ	*ẻ*	*Ẻ*
ओ	*o*	*O*
औ	*ỏ*	*Ỏ*
अं	*a˙*	*A˙*
अः	*a:*	*A:*
क्	*k*	*K*
ख्	*k̰*	*K̰*
ग्	*g*	*G*
घ्	*g̰*	*G̰*
ङ्	*n̰*	*N̰*
च्	*c*	*C*
छ्	*c̰*	*C̰*
ज्	*j*	*J*
झ्	*j̰*	*J̰*
ञ्	*n᪲*	*N᪲*
ट्	*t*	*T*
ठ्	*t̰*	*T̰*
ड्	*d*	*D*

D̓evanāgarī	**Roman Lower Case**	**Roman Upper Case**
ढ्	*d̰*	*D̰*
ण्	*n̥*	*N̥*
त्	*t̓*	*T̓*
थ्	*t̰̓*	*T̰̓*
द्	*d̓*	*D̓*
ध्	*d̰̓*	*D̰̓*
न्	*n*	*N*
प्	*p*	*P*
फ्	*p̰*	*P̰*
ब्	*b*	*B*
भ्	*b̰*	*B̰*
म्	*m*	*M*
य्	*y*	*Y*
र्	*r*	*R*
ल्	*l*	*L*
व्	*v*	*V*
श्	*s̰*	*S̰*
ष्	*s͚*	*S͚*
स्	*s*	*S*
ह्	*h*	*H*
ळ्	*l͚*	*L͚*
क	*ka*	*Ka*
ख	*k̰a*	*K̰a*
ग	*ga*	*Ga*
घ	*g̰a*	*G̰a*
ङ	*n̰a*	*N̰a*
च	*ca*	*Ca*
छ	*c̰a*	*C̰a*
ज	*ja*	*Ja*
झ	*j̰a*	*J̰a*

Dᷤevanāgarī	**Roman Lower Case**	**Roman Upper Case**
ञ	*n͚a*	*N͚a*
ट	*ta*	*Ta*
ठ	*t̰a*	*T̰a*
ड	*da*	*Da*
ढ	*d̰a*	*D̰a*
ण	*n̻a*	*N̻a*
त	*tᷤa*	*Tᷤa*
थ	*t̰ᷤa*	*T̰ᷤa*
द	*dᷤa*	*Dᷤa*
ध	*d̰ᷤa*	*D̰ᷤa*
न	*na*	*Na*
प	*pa*	*Pa*
फ	*p̰a*	*P̰a*
ब	*ba*	*Ba*
भ	*b̰a*	*B̰a*
म	*ma*	*Ma*
य	*ya*	*Ya*
र	*ra*	*Ra*
ल	*la*	*La*
व	*va*	*Va*
श	*s̰a*	*S̰a*
ष	*s͚a*	*S͚a*
स	*sa*	*Sa*
ह	*ha*	*Ha*
ळ	*l͚a*	*L͚a*
ा	*ā*	*Ā*
ि	*i*	*I*
ी	*ī*	*Ī*
ु	*u*	*U*
ू	*ū*	*Ū*

Devanāgarī	**Roman Lower Case**	**Roman Upper Case**
ृ	*r̰*	*R̰*
ॄ	*r̰̄*	*R̰̄*
ॢ	*l̰*	*L̰*
ॣ	*l̰̄*	*L̰̄*
े	*e*	*E*
ॆ	*ė*	*Ė*
ो	*o̯*	*O̯*
ॊ	*ȯ*	*Ȯ*
ं	˙	˙
ः	⁚	⁚
ँ	˘	˘

प्रकाशकस्य टिप्पणी

ऋग्वेदे १०.१५८ तमे सूक्ते चक्षुसौर्यः ऋषिः जगत् सम्यक् द्रष्टुम् अवगन्तुं च नेत्राभ्यां देवं सूर्यं प्रार्थयते। एतत् सुन्दरम् दुरवगमं च जगत् विशेषतया येन सह वयम् परिशीलकाः भूत्वा अविभाज्यसम्बन्धेन व्यवहरामः तद् अगवन्तुं अस्माकं तादृशी स्पष्टता आवश्यकी एव इति कथने न कोपि संशयः। यद्यपि अद्यतनवैज्ञानिकाः एतं क्लिष्टं जानन्ति जगत्सम्बद्धम् एतद् बहुमुखीनत्वम् अवगम्य बहिर्भूतान् विभिन्नान् परिणामशीलान् अंशान् स्वसिद्धान्तेषु अन्तर्भावयितुम् अविरतं प्रयतन्ते च तथापि अतिविस्तृतं क्षेत्रम् इतोप्यधिगन्तुम् विद्यते एव । मीमांसकाः अद्यतनवैज्ञानिकाः इव भौतिकास्तित्वे सम्यक् निरूढाः भवन्ति। ते लौकिकविज्ञानविषयेषु प्रत्यक्षे अनुमाने च दत्तबुद्धयः सन्ति तथापि धार्मिकांशेषु वेदस्य तेषाम् दृढस्य मीमांसस्य प्रभुत्वम् एव अङ्गीकुर्वन्ति वेदविधितकर्मकाण्डानुसारजीवनेन् धर्मसाक्षात्कारंकुर्वन्ति च। वस्तुतः अपूर्वम् इत्यस्य विषयस्य चर्चायां मीमांसकाः इदम् अभिप्रयन्ति यत् यस्य कस्यचिद् वेदविधितकार्यस्य प्रत्यक्षानुमानाभ्यां किमपि प्रयोजनं दर्शयितुं न शक्यते चेत् तदा एव अपूर्वं प्रयुज्यते इति। व्यवहारोपयोगिनि अस्मिन् सुरुचिरे लघुप्रबन्धे लेखकः तया दृष्ट्या वेदविज्ञानं प्रतिपादयति। अत्र कृतेः शीर्षिकायाम् उपयुक्तः षष्ठीविभक्तिप्रत्ययः स्पष्टम् अवगन्तव्यः। एषा शीर्षिका अस्ति अरुणप्रश्नस्य विज्ञानम् न तु अरुणप्रश्ने विज्ञानम्। अयं मीमांसकाणाम् एव दृष्टिकोणः नास्ति। विश्वे समग्रे एताम् एव दृष्टिं बह्व्यः स्थानीयाः धार्मिकव्यवस्थाः स्वकीयया पद्धत्या साक्षात्कारंकुर्वन्ति। अविदितेश्वरवादः प्रकृत्युपासकः अमेरिकास्वातन्त्र्योद्यमस्य पितृतुल्यः च थामस् पैनः The Age of Reason इतिनामके प्रसिद्धे स्वग्रन्थे उल्लिखति यत् **अस्माकं शास्त्रविज्ञानं सर्वम् अपि धर्मशास्त्रे एव व्युत्पत्तिं भवति। तत्शास्त्रविज्ञानातः एव अस्माकं सर्वाः अपि कलाः निर्गताः** इति। एतस्मिन् ग्रन्थे रचयितुः अयम् एव आशयः यत् मीमांसकाः अन्ये वेदपण्डिताः अद्यतनवैज्ञानिकाः तन्त्रज्ञाः च सम्भूय सर्वेषाम् मानवाणाम् श्रेयसाय वेदस्य विज्ञानम् अवगच्छेयुः इति। स्वतस्सिद्ध्या एषः विषयः अपि तादृशाय सहयोगाय कल्पते। अत्र महाभारतस्य ३.७.७२.८ तमे श्लोके बाहुकरूपेण स्थितस्य महान्नलमहाराजस्य पुरतः आत्मनः सङ्ख्याज्ञानस्य प्रदर्शनात् पूर्वं राज्ञा ऋतुपर्णेन उक्तं यत् तस्य उल्लेखनम् उचितं स्यात्। ऋतुपर्णः वदति यत् **सर्वः सर्वं न जानाति सर्वज्ञो नास्ति कश्चन। नैकत्र परिनिष्ठास्ति ज्ञानस्य पुरुषे क्वचित्॥** बहुशास्त्राणां युगपतध्ययनावश्यकतायाः अद्यतने काले इयं कृतिः शोधकर्तॄणां नूतनमार्गान्वेषणे नितराम् उपयोगाय भवति। अस्याः अक्लेशकरी शैली क्रमका विषयप्रस्तावनव्यवस्था च एतम् पुस्तकम् माध्यमिकशालातः महाविद्यालयपर्यन्तं विद्यार्थिभ्यः बहु कुतूहलाय कल्पते।

सामान्यजनेभ्यः वैदिककर्मकाण्डाणाम् अवगमनाणाम् दृढी कुर्वन् विविधाणाम् विज्ञानविषयाणाम् उपरि स्पष्टता किरणायते इयं कृतिः। उपसंहारे एकस्मिन उत्कृष्टलेखे आचार्यस्य अरविन्दस्य इमाम् उक्तिं स्मरामः। **सनातनधर्मस्य आधारः वेदाः इति विश्वसिमः। हिन्दुधर्मसिद्धान्ते इदं निगूढं दिव्यत्वम् इति मम भावना। परन्तु काचन अवगुण्ठनम् अपसारणीयम् काचन यवनिका उत्कर्षणीया। तत् ज्ञातुम् परिशोधितुं च शक्यम् इति विश्वसिमः। सन्यासाय न अपि तु जगत्यजनेषु जीवनाय तस्य पुनराविष्कारे प्रयोगे च भारतविश्वयोः भवितव्यता प्रतितिष्ठति इति विश्वसिमः।** इमां कृतिं भवन्तः आस्वादयन्ति भवतां वेदानां विज्ञानदृष्ट्या संशोधने प्रोत्साहयति सा च इति अस्ति अस्माकं आशा।

Publisher's Note

In *Ṛgveda* 10.158 *Sūkta, Ṛṣi Cakṣu Sorya* beseeches *Sūrya Devatā* for eyes to correctly see and understand our world. There is no doubt we need that kind of clarity for understanding our beautiful yet bewildering universe, especially when we as observers are interacting intimately with what we are observing. Scientists understand this complexity and are continuously attempting to internalize the external variables in their models, but there is still a vast ground to cover. Like the scientists, *Mīmā˙sakās* are firmly grounded in physical reality. They do accept the science's *Pratyakṣa* and *Anumāna*, but when it comes to *Ḍarma*, they rely only on their solid exegesis of the *Veda* and embody it through *Veda's* mandated ritualistic way of life. In fact, in the discussion on *Apūrva*, the *Mīmā˙sakās* say that only when *Pratyakṣa* and *Anumāna* do not show a physical benefit of an action enjoined by the *Veda* should one ascribe that to *Apūrva*. In this interesting succinct work of practical significance, the author outlines the science of the *Veda* along those lines. It is essential to note the preposition "of" in the title of the book. The title is "The Science of *Aruṇapraṣna:*" and not "The Science in *Aruṇapraṣna:*." This is not just the view of the *Mīmā˙sakās* alone. World over most ethnic-religious systems embody this perspective in their own unique way. Deist, nature worshipper, and the father of the American independence movement, Thomas Paine, in one of his most famous works, "The Age of Reason" said, "**It is from the study of the true theology that all our knowledge of science is derived, and it is from that knowledge that all the arts have originated.**" The author's goal in this work is to foster collaboration amongst the *Mīmā˙sakās*, other *Veda* scholars, scientists, and engineers to understand the science of *Veda* for everyone's benefit. In fact, the subject matter is such that it demands this kind of cooperation. It is helpful in this context to quote *Rājā Ṛtuparṇa*, from *Mahābārata* 3.7.72.8. *Ṛtuparṇa*, before demonstrating his statistical estimation skills says to *Bāhuka* his charioteer, who is none other than great *Nala Mahārāja*. "***Sarva: Sarvà Na Jānāti Sarvajṇo Nāsti Kaṣcana । Nekatra Parinisṭāsti Jṇānasya Puruṣe Kacit ॥*** " **Everyone does not know everything. There is no one who knows it all, and so complete knowledge is never established in any one person.** This book is an excellent resource for academicians looking for new avenues of research especially considering today's world of interdisciplinary studies. Its lucid style combined with structured presentation makes it a wonderful book for students of all ages from middle schoolers to graduate students. For the lay readers, it sheds clarity on various scientific concepts while offering a deeper understanding of the *Vedic* rituals. In closing, we remember what *Ācārya Arobindo* said in one of his great essays, "**I believe that Veda to be the foundation of the Sanatan Dharma; I believe it to be the concealed divinity within Hinduism,—but a veil has to be drawn aside, a curtain has to be lifted. I believe it to be knowable and discoverable. I believe the future of India and the world to depend on its discovery and on its application, not to the renunciation of life, but to life in the world and among men.**" We hope you enjoy this work and it inspires you to research the *Veda* scientifically.

Forewords

H.H Mounaswami
Founder

SRI SIDDHESWARI PEETAM

(Mouna Swamy Mutt)
COURTALLAM, TENKASI DISTRICT - 627802
TAMIL NADU

H.H Poojya Swamiji
Peetadhipati

Peetadhipati: **Paramahamsa, Parivraajakaachaarya, Jagadguru**
His Holiness Sri Sri Sri Siddheswarananda Bharathi Maha Swamiji

GENERAL SECRETARY	ORGANISING SECRETARY	CHIEF OPERATING OFFICER	P.R.O
P.S. Kantha Rao	Mataji Ramyaananda Bharathi Swamini	Mocherla Venkata Lakshmi Sashibhushan	Purnajnanananda Saraswati Swamiji

आशिर्वादश्रीमुखम्

Aruṇapraṣna is housed in *Kṛṣṇayajurveda Tettirīya Ṣāka Āraṇyakam*. It is recited for the benefits of disease-free body, *Ārogyadṛḍagātra* (आरोग्यदृढगात्रं), life enhancement, *Āyuṣābivṛddi* (आयुषाभिवृद्धि) and banishment of premature death, *Apamṛtyudoṣanāṣana* (अपमृत्युदोषनाशनं) in addition to numerous other benefits.

Sri Sitaram Ayyagari has brought out the book named "Science of *Aruṇapraṣna*". It highlights information encoded in the *Mantras* of *Aruṇapraṣna*. He developed a framework to interpret the *Vedic Mantras* in a scientific manner that was rational while maintaining the sanctity of interpretations. The book deals with wide range of scientific topics including: the creation of our planet and its climate, astronomy, quantum physics, etc. He classified the *Vedic Mantras* from *Aruṇapraṣna* in both their scientific and technological capacities into six chapters, namely *Būmi, Vāyumaṇdalam, Varṣa, Gurutvākarṣaṇa, Sūryaraṣmi,* and *Yajṇenabandu.*

May the divine bless Sri Sitaram to bring out many more works and inspire the next generation of students, *Veda Panditas*, scholars, and academics to continue to find innovative ways to analyze the *Vedas* from a scientific aspect and bring out its knowledge to better understand the universe.

Dr. Remella Avadhanulu

Chief Executive, Shri Veda Bharathi.

Deputy Director Computers, (Retd.), Nizam's Institute of Medical Sciences, Hyderabad.

M.Sc.(Nuclear physics), M.A.,Ph.D.(Sanskrit), M.A.,Ph.D.(Jyotisha), D.Litt(Hon).

https://www.shrivedabharathi.org

The *Vedas* are the sources of all knowledge and are universal in their application. They are useful for the moral, spiritual, and physical guidance and uplift of humanity. Recent findings world over confirm the relevance of *Vedas* even today for both spiritual practice and scientific research. A close study of the *Vedic* hymns with the help of the *Ṣāśtras* in general and *Niṛuktá*, *Vyākaraṇa*, and *Mīmāṃsa* in particular help to understand the *Vedic* literature in its totality and to present the concepts and contents of modern sciences in their true sense.

The *Vedas* are like deep oceans with hidden jewels. Several sages, saints, scientists, and scholars have dived deep and could bring into light invaluable gems of universal knowledge for the benefit of the mankind. It is observed that the subjects are covered either in an elaborate form, conceptual form, in a seed form, allegorical form, or story form. The scientific knowledge in ancient *Ḇāratam* is illustrated with few examples given below.

Cause & Effect Theory of Science

Just as it was in ancient times, the science of physics today is based on the cause and effect theory. This principle is observed in *Veṣeṣikadarṣana* as follows:

कारणाभावात् कार्याभावः ।

न तु कार्य्याभावात् कारणाभावः ।

Kāraṇābāvāt kāryābāva: ।
Na tu kāryyābāvāt Kāraṇābāva: ।।

No effect if there is no cause. But absence of effect does not mean absence of cause.
(*Vėṣeṣika* 1.2.1, 1.2.2)

Conservation of Mass and Energy

The concept of conservation of mass and energy can be traced in the famous verse of *Gītā*:

नैनं छिन्दन्ति शस्त्राणि नैनं दहति पावकः ।

न चैनं क्लेदयन्त्यापो न शोषयति मारुतः ॥

Nėna· çindanti ṣastrāṇi Nėna· dahati pāvaka: ।
Na cėna· kledayantyāpo Na ṣoṣayati māruta: ।।

It (the body-holder) cannot be pierced by weapons nor can fire burn it. Water cannot wet it and winds cannot dry it either. (*Gītā* 2-23)

Theory of Gravitation, Concept of Spherical Shape of Earth & Scientific Concept of Eclipses

It is generally stated that Newton, belonging to 17th century AD, was the first scientist to propose that earth has the power to attract all objects towards it due to its gravitational force. But a cursory perusal of our ancient literature brings out stunning information on this topic. *Varāhamihira*, the great astronomer who lived in India in 6th century AD, recorded in his *Paṇcasiddāntikā*, that all the objects in the universe attract each other.

He also mentions the spherical shape of earth.

पञ्च महाभूतमयस्तारागनपञ्जरे महीगोलः ।

क्षेयस्कान्तस्थो लोह इवावस्थितो वृत्तः ॥

Paṇca mahābūtamayastārāganapanjare mahīgola: ।
Kṣeyaskāntasto loha ivāvastito vṛtta: ।।

In the group of great celestial bodies, the spherical shaped earth, made of *pañcamahābhūtas*, exists in space, like a piece of iron attracted by a magnet, from all the sides.

In fact the same concept was further declared by *Bāskarācārya* of 12th century as follows:

आकृषिशक्तिश्च महीयत् खस्तम् गुरु स्वाभिमुखम् स्वशक्ति ।

आकश्र्यते ततप्ततीव भाति समे समम्तात् क्व पतत्ययम् खे ॥

Ākṛṣiṣaktiṣca mahīyat kaśtam guru svābimukam svaṣakti ।
Ākarṣyate tataptatīva bāti same samamtāt kva patatyayam ke ॥

This earth attracts whatever solid materials are in the space, by her own force of attraction towards her (earth). All those subjected to this attractional force fall, to the earth. Due to equal force of attraction among the celestial bodies, where can each among them fall? (*Siddāntaṣiromaṇi* - *Buvanākoṣa* 19–6)

A reference was made in *Nyāyakandalī* of *Śrīdara* wherein the theory of gravity was clearly mentioned:

गुरुत्वं जलभूम्यो: पतनकर्मकारणम् ।

अप्रत्यक्षं पतनकर्मानुमेयं ... ।।

अथावयवानां गुरुत्वादेव तस्य पतनं तदवयवानामपि स्वावयवगुरुत्वात् पतनमिति सर्वत्र कार्ये तदच्चेद: ।

Gurutva jalabūmyo: patanakarmakāraṇam ।
Aprtyakṣa patanakarmānumeya ... ॥
Atāvayavāna gurutvādeva Tasya patana
Tadavayavānāmapi svāvayavagurutvāt patanamiti
Sarvatra kārye taducceda: ।

Gravity is the cause for falling of liquids and solids. It is invisible and is inferred by the falling motion. Gravity acts not only on the body, but equally on its finer constituents.

Thus the ancient literature of India containing several scientific concepts in an embedded form is a grand source of research, from long time, at several places, by several scientists. In the present context, the "*Aruṇapraṣna*", the source of the present investigation, is the first chapter of the section titled "*Āraṇyaka*" of *Tettirīya* branch of *Yajurveda*. This has been the source of research from long time. In the recent history, Dr. Rani Ramakrishna, who was subsequently renamed as Swami Tattvavidananda Saraswati after renunciation, and also who is spearheading the cause of Vedic philosophy with scientific approach through out the globe, published his research findings in a book titled "Science in Krishna Yajurveda" in the year 1994. The book was critically reviewed and appreciated by Mahamahopadhyaya Professor Pullela SriRamachandrudu, Head of the Department of Sanskrit, Osmania University, Hyderabad. Similarly Dr. Remella Avadhanulu, who was hailed and awarded by the Government of India, in his book titled "Science and Technology in Vedas and Sastras", published in 2007, made extensive references to different textual portions of *Aruṇapraṣna*, while bringing out several hidden scientific aspects.

The author of this book Sri Ayyagari Sitaram qualified himself adequately through acquisition of knowledge of modern sciences initially and of *Vedic* texts and their commentaries subsequently. After necessary preparation, he dedicated himself to the cause of deciphering the *Aruṇapraṣna*, with reference to formation of earth, environment, cosmos etc., in the background of theoretical and applied research findings of the modern sciences. Accordingly the *Vedic mantras* are critically analysed for scientific and technological aspects and the findings are broadly distributed into different groups, and are covered in the following chapters:

Būmī

Vāyumaṇdalam

Varṣā

Gurutvākarṣaṇa

Sūrya Raṣmi

Yajṇenabanḍu.

Every chapter is enriched with the texts of the *mantras*, and the findings are furnished in the form of hypotheses. The presentation is in a simple language with straight forward description of the *Vedic mantra* texts on one side and contemporary knowledge on the other side, along with necessary illustrations on all important concepts.

We are sure that this work will inspire several scientists and scholars, to go simultaneously for a fairly high level of understanding of modern sciences, and also of ancient Indian knowledge systems, for continuation of this kind of research, and to bring out convincing and converging propositions in future.

Congratulating the author for his extraordinary work,

With best wishes

Dr. Remella Avadhanulu

Hyderabad

7th January, 2022

Dr. ChandraSekhar Tewary

Assitant Professor, Metallurgical and Materials Engineering,
Principal Investigator, Nano Engineering Research Group,
Indian Institute of Technology Kharagpur.
Ph.D., Indian Institute of Sciences, Bangalore.
http://www.iitkgp.ac.in/department/MT/faculty/mt-chandra.tiwary

The Science of *Aruṇapraṣna* by Sitaram Ayyagari is well written and nicely documented book. A perfect balance of both quality and quantity. In the era of interdisciplinary work, the boundary of individual field has diffused. All the work of current century clearly shows that interrelationship between different scientific areas. At the same time, as science is progressing and providing or strengthening society with newer tools all the way from high-end microscope to artificial intelligence and machine learning based technology, today we can understand the unknown and unexplored areas. In such an exciting time, it is very interesting and exciting to revisit few of the traditional practices such as *Aruṇapraṣna* from a new and exciting angle as explained in the book.

I would like to Congratulate Sitaram ji for putting such nice content. He has made a successful initial attempt to correlate two different areas. I am sure it will reach to broader readers and young minds will take it up further.

With best wishes and regards

Dr. Chandra Sekhar Tiwary
22nd March, 2022

Brahmasri Samavedam Shanmukha Sarma

Founder Rushipeetham

Purāṇa, Itihāsa, and *Veda Pravacanakarta,* Publisher and Telivision Personality

https://rushipeetham.org/

The term *Veda* itself means *Jñāna.* It is a treasure of wisdom that reveals truths, which cannot be realized through observation or inference. The Westerners have established the *Vedas* as a mere religious text or an ardent expression of the worshippers of natural forces. However, in the past few decades, especially during the 20th century, a section of intellectuals well-versed in the *Bāratīya Vedic* philosophy and also having the capability to challenge the established incorrect notions of the Westerners have emerged.

When the *Vedas* are researched with a scientific perspective, the truth that reveals the secrets of the infinite universe is visualized. Scientists are welcoming such ideas with an open mind. This book is one among the several wonderful works that have come out in this conducive environment.

This book has been authoritatively written based on the thorough study of the latest scientific discoveries from Cosmology to Quantum Physics and a wholistic research of every word of the *Aruṇapraṣna* as per the appropriate *Bāṣyas*.

The uniqueness of this text lies in interpreting the meaning of the words in agreement with science, without contradicting the traditional meaning of the *mantra*, and leaving no scope for allusions or misinterpretations.

The effort, research, thought, and interpretation of Sri Ayyagari Sitaram garu in authoring this text are highly commendable. It will be clearly evident to the readers that the book is neither opiniated nor creatively imagined. The style of analysis and explanation of several established scientific concepts and visualization of the same within the layered meanings propounded in the *Vedic mantras* is mesmerizing. The writer has proved that *Aruṇapraṣna* has expounded not only solar science but also creation and evolution of the universe and emergence of the *Paṇcaḇūtas*.

Veda Sūktas are eternal complex truths visualized by undefinable extra-sensory knowledge, which blossomed through *Tapasyā*. Those who possessed such knowledge were *Ṛṣis*. Knowledge gleaned only from sensory perceptions is limited. Therefore, it is not possible for immature minds to grasp this higher knowledge. Great minds have interpreted these *Sūktas* as per their intellectual capacities. However, the *Sūktas* have hidden knowledge beyond all these interpretations.

The author studied the *Veda Sūktas* and the related *Ḇāṣyas* on one hand and also studied the established and emerging related Science on the other hand as a *Tapasyā*. The comparative analysis between them in this text is profound. Several scientists and *Vedic* scholars have already commended his work.

I have had numerous opportunities to discuss with the author during the development of this work and now I have the pleasure of reviewing this wonderful completed work. The method in which the various topics, creation and evolution of earth, explanation of the atmosphere, rain, gravitation, sunlight, and the Science of *Yajṇa* have been arranged in the text is stupendous. This book is a must read for the present generation, which requires rational justification. It is praiseworthy that the presentation style of the book is in line with style employed by the current scientific works.

In addition to the satisfaction of having obtained the *Vedic* knowledge as an ancestral inheritance, I have the joy as a global citizen that the *Sanātana Ḍarmā* knowledge is useful for the progress of all mankind. While focusing on the *Ādiḇōtika* interpretation, the text also touches upon the

Ādidévika and *Ādyātmika* interpretations as well.

In this age of technologically expanding globalization, works such as this book help widely disseminate the secrets of the *Vedic* knowledge.

The ancient *Veda* is not in contradiction with science. Works such as this book help us realize that *Veda* not only explains the science of the material world but also expounds the metaphysical, *Paramārṭatatvā* and thus going beyond "science", the *Veda* shines as a "super-science".

Sri Sitaram garu is laudable for his dedication to the discovery of truth and successfully bringing out this work. May *Rāma* bestow his blessings on him.

Buḍa Jana Viḍeya

Sama Veda Shanmukha Sharma

21st December, 2021

Organization of the Book

This book is organized into seven chapters, one for each of the six scientific topics discussed and one chapter for the Introduction. Each chapter other than the Introduction chapter, has the following sections.

1. Science – The existing science on the chapter's topic.
2. *Aruṇapraṣna* – Summary discussion of the topic in *Aruṇapraṣna* along with the related *Aruṇapraṣna Ma'tra's* and their meanings.
3. Postulates – Author's three postulates on the topic.
4. Postulate Discussion – Detailed discussion of each of the three postulates is presented in three separate sections, one for each postulate.
5. Conclusion – Summary of the postulates and suggested topics for future research.

Two separate glossaries, one for scientific terms and the other for Sanskrit terms have been provided in the back matter of the book. Also, the citations to scientific papers notated in the book are references to the items in the bibliography. Readers are encouraged to refer to the glossaries and the bibliography while reading the book to get a better understanding of the topic. The bibliography has been organized into seven sections corresponding to each of the seven chapters. Specifically, the Introduction section of the bibliography includes a list of works that provide the foundation for understanding the *Aruṇapraṣna.* These works are useful for all the chapters in the book but to avoid repetition within the bibliography they are only mentioned in the Introduction section of the bibliography. In addition, the entire *Aruṇapraṣna* text in *Devanāgarī* along with *Svarā* notation has also been provided in the back matter. Finally, following normal publishing guidelines for a work of this kind, an index has been provided at the end of the book.

Chapter 1
Introduction

Introduction

In the last 75 years or so, scientists and traditional *Veda* (वेद) scholars have done valuable work in their various personal capacities to explore the *Veda* from a scientific perspective. Several scholars at Banaras Hindu University's Sanskrit Vidya Dharma Vignan Sankay, Maharishi Sandipani Veda Vidya Pratishthan, various Rashtriya Sanskrit Sansthans, Kapila Sastry Institute of Vedic Culture, Bhandarkar Oriental Research Institute, Kuppuswamy Sastri Research Institute, Vaidika Samsodhana Mandali, Shri Veda Bharati, and many other institutions across Bharat have made significant contributions in this direction. However, due to various known and unknown reasons, genuine collaborative research between science and *Veda* has not been born yet, or at best, is in its early infancy. It indeed has not been enough to be categorized as a serious subject matter by the scientific community. What little has been done is mainly related to the ancillary disciplines such as *Yoga* (योग), *Āyurveda* (आयुर्वेद), *Vāstuśāstra* (वास्तुशास्त्र), and *Bāratīya* Astrology. Moreover, the work has remained an intellectual discussion within the circle of specialized research scholars and has not yet made its way into the educational curriculum. In the West, universities and other institutions have virtually cloistered themselves into the narrow confines of what they call Indology or South Asian studies. They are so far removed from this line of thinking that nothing can be expected from them anytime in the near future. Both in *Bāratam* (भारतम्) and in the West, the science side of the house does not recognize this potential despite the ubiquitous emergence of interdisciplinary studies. This compact book is a humble but bold attempt to think in that direction and encourage collaboration amongst traditional *Veda* scholars, *Karmakāṇḍa* (कर्मकाण्ड) practitioners, scientists, and engineers to build this much-needed research discipline.

There is a misconception that *Vedic* (वेद) rituals and life in general are two separate things. This is because our current mechanical way of life, which is devoid of *Vedic* rituals, is divorced from cycles of nature. Unsurprisingly, this leads one to an incorrect understanding of the *Veda* because this unhealthy lifestyle isolates *Veda* into the category of spiritual pursuits, thus removing it from everyday life. To gain a scientific understanding of the *Veda*, one must first understand *Vedic* rituals. More importantly, only through the practice of *Vedic* rituals will *Veda* be able to manifest in our daily lives physically. This manifestation enables us to experience the science inherent in the *Veda* empirically.

The *Vedic* way of life and the accompanying mandatory rituals follow specific cycles of nature. These are the diurnal solar cycle, the monthly lunar cycle, the seasonal solar cycle, and the biological life cycle of a person. Underlying all these cycles is *Ṛtam* (ऋतम्), the *Vedic* word for the way the universe operates. Even the optional *Vedic* rituals that are performed for specified benefits are based on an understanding of *Ṛtam*. The fundamental *Vedic* assumption is that the best way to live is to stay in sync with *Ṛtam*.

Living beings other than humans are instinctually in sync with nature. On the other hand, humans depend more on their mental faculties than their instincts to guide their way of life. More often than not, this leads us to fall out of balance with mother nature. Modern society's environmental, health, and various social problems are examples of this imbalance. However, to live in sync with nature without devolving into an instinct-driven animal is not an easy task. It requires a very carefully designed lifestyle. This kind of structured life becomes necessary because human societies are too complex to rely solely on animal instincts.

Veda is the scientific insight of our *Ṛṣis* (ऋषि) into the cycles of *Ṛtam*. They developed a *Vedic* rituals-based lifestyle, which allows us to live in sync with nature. The common understanding of the *Veda* is limited to the *Saṁhitā* (संहिता) and *Upaniṣad* (उपनिषद) portions. Unfortunately, this excludes the entire *Vedic* ritualistic complex in the *Brāhmaṇam* (ब्राह्मणम्) and *Āraṇyakam* (आरण्यकम्) portions of the *Veda*. The rituals explained and mandated in those portions enable one to live in accordance with *Ṛtam*. These rituals are systematized and described in greater detail in the *Kalpasūtras* (कल्पसूत्राः) of the *Vedāṅga* (वेदाङ्ग). They are finally developed into implementation manuals in texts such as the *Prayogas* (प्रयोगाः), *Paddatis* (पद्धतयः), etcetera.

Science is the observation and understanding of how nature works.

We all know how science translates into our everyday lives, directly and indirectly, so there is no need to elaborate further. The *Vedic Ṛṣis* had their unique way of observing the universe. They are called *Draṣṭas* (द्रष्ट), the ones who perceived nature wholistically. In particular, they studied the link and interaction between the cycles of mother nature and human lifestyles. They then applied that wisdom to develop a sustainable and fulfilling way of life at the individual and societal level. With advances in Quantum Physics and the Cognitive Sciences, scientists are at the periphery of this highly developed, integrated thought of the *Vedic* seers. In this small work, we are attempting to uncover the scientific understanding behind one of the *Vedic* rituals, the *Aruṇaketukāgni Srauta Yajṇa* (अरुणकेतुकाग्नि श्रौत यज्ञ), described in the first *Prapāṭaka* (प्रपाठक) of the *Tettirīya Āraṇyakam* (तैत्तिरीय आरण्यकम्) in the *Kṛṣṇayajurveda* (कृष्णयजुर्वेद). This *Prapāṭaka* is commonly known as *Aruṇapraṣna* (अरुणप्रश्न).

Aruṇapraṣna is an excellent place to begin the pursuit of a scientific understanding of the *Veda*. It discusses various interconnected topics such as earth's formation, earth's atmospheric layers, solar electromagnetic radiation, time, gravity, seasons, and rain. The science inherent in the metaphorical representations in *Aruṇapraṣna's* discussion of these topics is quite evident. If there are any doubts about the underlying science here, they can be quickly dispelled with a study of the *Sāyaṇācārya's Bāṣya* (सायणाचार्यभाष्य) of the *Aruṇapraṣna*. That being said, since the *Vedic* way of life is a strongly orthopraxical system, understanding the natural world is necessary but not sufficient for it. A *Vedic* lifestyle mandates living in sync with nature through prescribed rituals. Therefore, *Aruṇapraṣna's* central purpose is the *Aruṇaketukāgni Yajṇa* ritual.

This work will discuss selected scientific topics from the perspective of *Aruṇapraṣna*. The purpose of this book is not to show concordance between science and *Aruṇapraṣna* but to present *Aruṇapraṣna*'s view on these scientific topics. Science is open-ended. Within the frameworks of assumptions, constraints, and measurements, phenomena are described. Expanding the realms of understanding is what makes science progress by process of repeatability, falsifiability, and consensus. That being said, not all scientific concepts have the same degree of consensus amongst the scientific community. For example, the current understanding of Newton's Laws of Motion or Einstein's Photoelectric Effect is far more stable than that of how the earth was formed or how condensation happens for rainfall. So, science is still far from having a conclusive understanding of these types of concepts, and even within the scientific community, there are competing hypotheses on these subjects.

The epistemology of the *Vedic* philosophical tradition is based on six sources of knowledge, known as *Pramāṇas* (प्रमाणाः). Namely, *Pratyakṣa* (प्रत्यक्ष) meaning sensory or extended sensory perceptions, *Anumāna* (अनुमान) meaning reasoning, *Upamāna* (उपमान) meaning comparison and analogy, *Ṣabda* (शब्द) meaning testimony of past experts, *Artāpatti* (अर्थापत्ति) meaning circumstantial implication, and *Anupalabdi* (अनुपलब्धि) meaning negative perception. As per this framework, science is based on *Pratyakṣa* and *Anumāṇa* only. *Pūrvamīmā˙sā* (पूर्वमीमांसा) is the philosophical exegesis of the *Veda*, and more specifically, of the *Vedic* ritual. It admits all the six *Pramāṇas* to prove its position but ranks *Ṣabda* as the primary and only reliable source of knowledge. For *Mīmā˙sakas* (मीमांसकाः), *Veda* being *Anādi* (अनादि), timeless, *Aporuṣeya* (अपौरुषेय), unauthored by anyone, and valid in and of itself, *Svatahapramāṇya* (स्वतह्प्रमाण्य), it is the most important and infallible *Ṣabda*. This book is not a place for a detailed discussion on these concepts, but it is essential to have a basic understanding of the sources of knowledge in the *Vedic* tradition. Interestingly, *Aruṇapraṣna* 1.2.1 outlines its *Pramāṇas*, and based on *Sāyaṇācārya Bāṣya* on the same, includes a combination of *Pratyakṣa*, *Anumāna*, and *Ṣabda*. *Ṣabda* for *Sāyaṇācārya* consists of the *Veda* or *Ṣruti* (श्रुति), and *Smṛti* (स्मृति), which includes *Purāṇas* (पुराणाः), *Itihāsas* (इतिहासौ), and other related ancillary sources.

In this book, I have presented my views as my hypotheses on these various topics. I want to emphasize that my hypotheses are my interpretations only, and not necessarily those of the *Veda*. I have made every attempt to ensure that my views are not in contradiction to the *Veda*. If there is a contradiction between my interpretation and *Veda*, I stand corrected because the *Veda* is infallible. Similarly, I have made every effort not to go against established scientific principles, but if it has happened, I stand corrected again. If, however, there is a contradiction between

Science and *Veda*, I am with the *Mīmā'sakas*, who urge further study and research to resolve this contradiction because the *Veda* is infallible.

A hypothesis becomes an accepted scientific principle only if there is a rational explanation based on existing scientific principles and if the hypothesis is proved by empirical evidence. In the Hindu philosophical tradition, this concept is explained under the discussion of *Anumāna Pramāṇa*. *Pratijṇā* (प्रतिज्ञा), hypothesis, becomes *Satya* (सत्य), Truth only when there is *Hetu* (हेतु), a rational explanation based on existing truths, and *Ḍṛṣṭānta* (दृष्टान्त), empirical evidence. In this book, I only state my hypotheses based on *Aruṇapraṣna* and discuss these hypotheses in some detail. I provide as much rational support for the same as is within my capacity. For each scientific concept, I first present the existing science on the subject. Then, I present three hypotheses based on my interpretation of *Aruṇapraṣna* on that subject. Finally, I discuss each of those hypotheses in some detail. Solid rational arguments and empirical evidence meant to substantiate my claims are major undertakings that should be taken up by those who have the necessary knowledge and resources for such an endeavor. The book essentially is my hypotheses on various scientific topics based on my interpretation of *Aruṇapraṣna*.

We start our scientific journey through *Aruṇapraṣna* with the chapter titled *Ḇūmī* (भूमि), which is a discussion on the formation of the earth. This chapter on the cosmology of our planet is then followed by *Vāyumaṇḍalam* (वायुमण्डलम्), a description, and analysis of the earth's atmospheric layers. We then delve into the chapter *Varṣā* (वर्षा) that talks about the fascinating natural phenomenon of rain. The following chapter, *Gurutvākarṣaṇa* (गुरुत्वाकर्षण) grounds us back on earth with an exploration of the subject of gravity. Chapter 6, *Sūryaraṣmi* (सूर्यरश्मि) is devoted to solar electromagnetic radiation. The final chapter, *Yajṇenabandu* (यज्ञेनबन्धु) presents my hypotheses on the Quantum Physics of *Yajṇa*. Each of my hypotheses offers an excellent opportunity for interdisciplinary research among scientists, engineers, *Veda* scholars, and *Vedic* ritual practitioners. I sincerely hope that this book provokes an interest in the subject in its readers and encourages them to study further and research.

Chapter 2
B̰ūmī

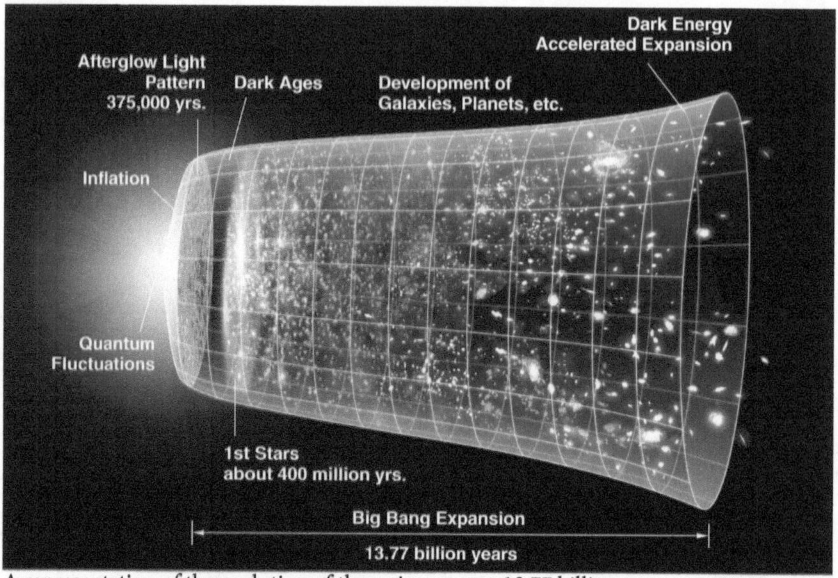

A representation of the evolution of the universe over 13.77 billion years.

NASA/WMAP Science Team, Public domain, via Wikimedia Commons.

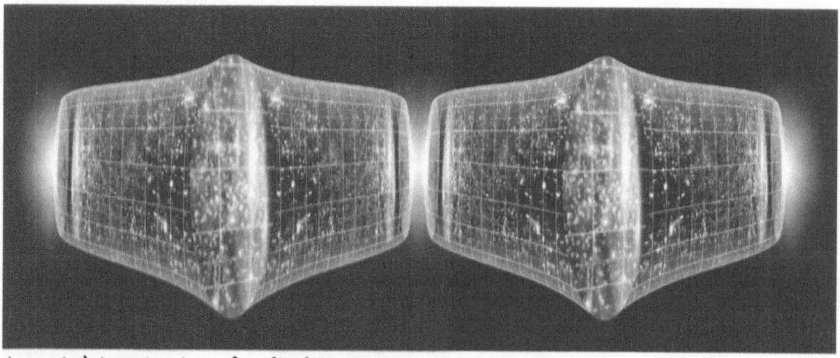

An artist's imagination of cyclical Big Bounce type universe.

Developed using first image on this page above from NASA/WMAP Science Team, Public domain, via Wikimedia Commons.

Science

The origin of the universe and the formation of our planet earth, as we see it today, has been a subject of speculation since time immemorial. Despite today's advanced observation techniques and analytical rigor, existing cosmological scientific theories are still hypotheses, and conceptually speaking are not much better than various mythological creation stories from around the world. Every earth formation theory, has hypotheses that are yet to be explained, and/or has contradicting observations. In many cases, when one of the unknowns or contradictions is resolved, new ones take its place. This is not surprising given the scale of time and space involved in this endeavor and our limited ability to observe, collect, and analyze information under these constraints.

Despite several theoretical and observed problems, the Big Bang theory, and its associated Lambda cold dark matter parametrization is the most accepted cosmological model of the universe. This model estimates the universe to be 13.8 billion years old. The main problem with the Big Bang theory is its inherent contradiction. On one hand, it relies on the laws of physics to explain the origin and development of the universe, but on the other hand, it assumes that all laws of physics breakdown in its fundamental assumption of what existed before the Big Bang: the infinitely dense Singularity. More recently, the theory and the concept of a constantly expanding universe has been challenged by cyclical concepts such as the Big Bounce, Ekpyrotic, and Quantum Physics based theories that postulate an eternal universe. Even the formation of the solar system, and our planet earth is unclear. The accepted model for the formation of the earth is the nebular hypothesis, which states that our solar system started in a solar nebula, which was the interstellar gas and dust created by the Big Bang. An external event initiated the gravitational collapse of this nebula. This collapse subsequently led to the formation of a protoplanetary disk. The center of the disk further collapsed to form the sun, and the rest of the disk formed other parts of the solar system, including earth.

An important concept which is useful in understanding this cosmological

Artist's concept of a protoplanetary disk, where particles of dust and grit collide and accrete forming planets or asteroids.

Pat Rawlings, NASA, Public domain, via Wikimedia Commons

model of the universe is Nucleogenesis. This is an attempt to explain the origin of the elements found both on earth and throughout the universe using nuclear physics. The theory states that at the onset of the Big Bang, the high temperature and high-density conditions produced quarks and gluons, the fundamental particles of matter. Through their interactions, hydrogen atoms emerged. The interaction amongst the hydrogen nuclei produced Helium, Lithium and Beryllium and two isotopes of Hydrogen, Deuterium and Tritium. Beryllium and Tritium were unstable and eventually degraded into Helium and Lithium as the universe cooled. Remarkably, all this happened within 20 minutes of the Big Bang. The remaining higher periodic elements from Carbon to Iron are believed to have been produced via thermonuclear fusion during various stages of stellar evolution across the universe. All other non-manmade heavier elements are considered to be products of supernovae. Some of the lighter elements are thought to also have been produced by cosmic ray fission processes.

Now coming back to the creation of the earth, almost all branches of the natural sciences have contributed to developing the story of earth's creation. Earth scientists' most valuable technique to understand the story is the radiometric dating of rocks. Based on this, they believe that the earth was formed approximately 4.5 billion years ago. The portion of the protoplanetary disk that did not collapse into forming the sun moved about in the space around the sun, under the influence of sun's gravity. Electrostatic forces led to the coalescing of the particles into larger sized particles. From causes yet to be understood, there was progressive accretion through collision of the particles into larger objects, which eventually led to the emergence of very large bodies that exerted gravitational forces, which in turn greatly accelerated their growth. Eventually, there emerged winners in these gravity driven collisions, and earth was one of them. The others were the remaining planets. Scientists claim that the asteroids in our solar system are the remaining portion of the protoplanetary disk that did not form any planet sized object.

All those collisions combined with the radioactive energy from some of the colliding rocks raised the temperatures to the point where all the rocky material that constituted the earth melted. Due to gravitational forces, the

Backjet of a drop of water after impact on a water-surface.

molten mass of rocky material separated into layers. The densest part of the molten mix settled to the center of the earth forming the core, the material with lesser density formed the mantle and the middle layer and remainder formed the earth's crust. The outer layer is called the crust because as the earth cooled the outer layer crusted over the semi viscous mantle beneath it. This viscous, hot mantle periodically erupted through volcanos and spewed dust and hot gases, which formed the earth's initial atmosphere. The next chapter in earth's story was the emergence of life, which needed water.

The origin of water and oceans on earth is unknown. Most scientists believe that all water came via icy planetesimals that collided with earth, but they have no clarity over when this happened. Some say that water came trapped within the rocky meteorites that collided to form the earth. Other scientists have concluded that more than half of the water on earth is older than sun itself and has come to us from outside of our solar system. In any case, it is believed that most of the water turned into water vapor during earth's hot and molten state and eventually came down as rain when the earth cooled. As more complex states of water are researched, it may also be possible that water existed within the molten layers of earth in some yet to be discovered state.

Scientists take the existence of water as a basis and go into abiogenesis to explain how simple cellular life originated on earth. It is believed that cellular life was born from inorganic matter, and then evolved into more complex forms. At some point in this miraculous development, cyanobacteria came into being and is thought to have produced the first large quantities of oxygen in the atmosphere. Oxygen led to the formation of the ozone protective layer and among other things supported the development of more complex life forms. Scientists rely on the Theory of Evolution to explain this process. Similar to abiogenesis, the Theory of Evolution is a hypothesis. Going back to abiogenesis, no laboratory experiment has yet been able to develop even a single cell organism from inorganic materials. The only thing that has been created in the lab are organic chemicals such as amino acids. Moving on to the Theory of Evolution and other inter-specie evolution theories, while various degrees of similarity in the genetic profile of various life forms on Earth are used

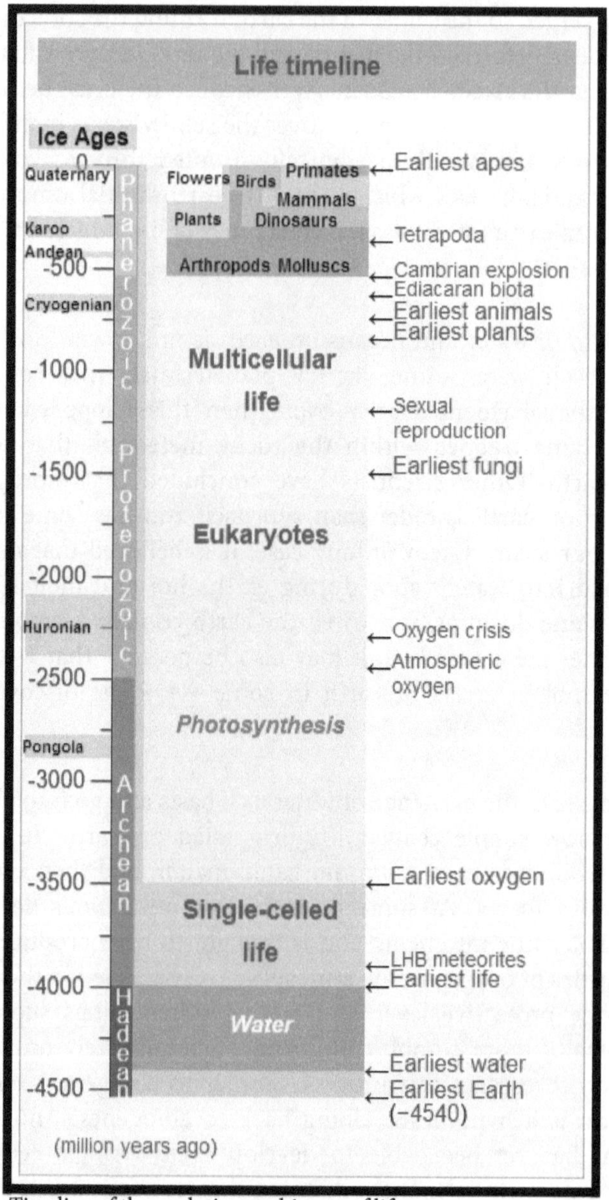

Timeline of the evolutionary history of life.

as a basis for macro evolution across species, these theories are mostly an extrapolation of an individual specie's evolution and the claim is that an entity as complicated as the human being has evolved progressively from an abiogenetic single cell organism. In the west and particularly in the United States, the evolution debate has unfortunately taken on a religious and political dimension. One is forced to pick sides in this beliefs debate and there is no space for a true scientific questioning of either abiogenesis or the Theory of Evolution in the public space. In addition to Abiogenesis, there are at least three other theories for origin of life namely, Panspermia Theory, Hydrothermal Vent theory and Primordial Soup theory.

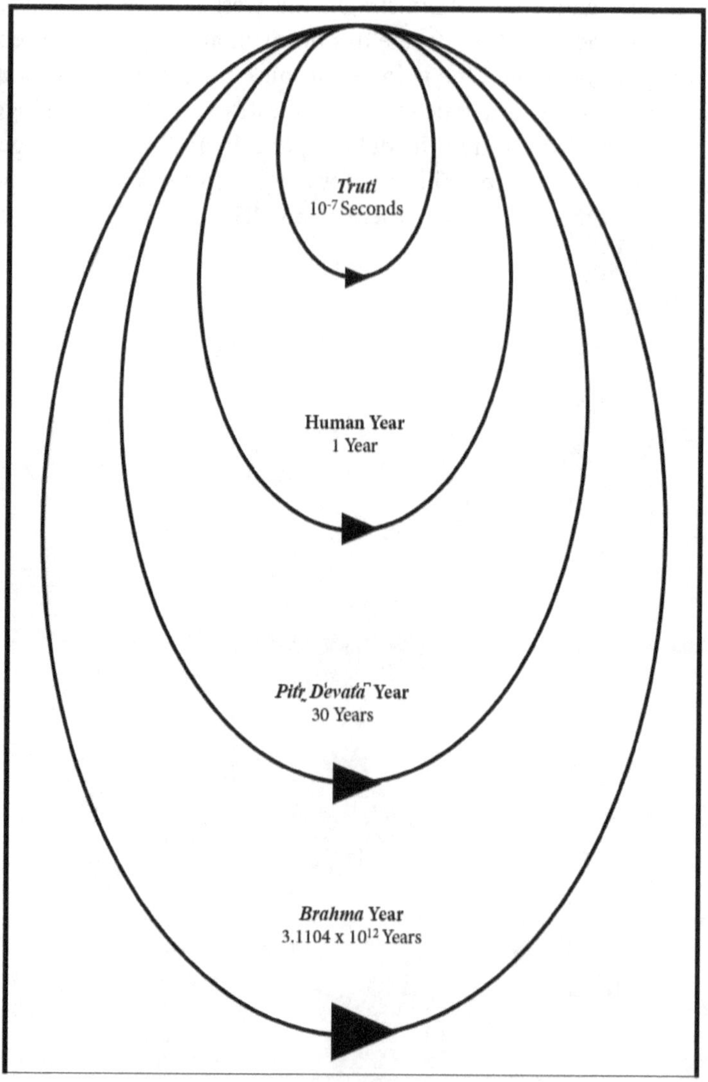

Hindu Cosmology is based on infinite cycles of time.

Aruṉapraṣna

The *Vedic* tradition has various theories on this subject. There is a complex relationship amongst these theories, which may even appear contradictory with one another. The tradition allows for this flexibility, and in fact, nurtures it. It accommodates theistic, agnostic, and interestingly even the apathetic approaches to this subject. Given the uncertainty on how the greater universe and our earth with all its life forms, came to be, this openness provides the scope for continuous learning, rather than getting locked into a dogmatic belief. That being said, *Vedic* cosmologies have some things in common. They are grounded in the belief that creation is cyclical with cycles taking trillions of years. They believe that human beings, animals and plants have always existed in all these endless cycles of creation. The creator is depicted as pure consciousness in the form of *Brahmaṉa* (ब्रह्मण), *Indra* (इन्द्र), *Prajāpati* (प्रजापति), *Tvaṣṭr* (त्वष्ट्र), *Vāgambṟṉī* (वागम्भृणी), *Viṣṉu* (विष्णु), *Rudra* (रुद्र), or the agnostic unknown. It is interesting to note that *Pūrvamīmā'sā*, probably the oldest of the six main philosophies of Hinduism, states that the existence of an omnipotent creator cannot be proven. At the end of this section of the chapter, I have presented the *Sūkta* 10.129 (सूक्त १०.१२९) from *Ṟgveda* (ऋग्वेद). In this *Sūkta*, *Veda's* creation postulates and at the same time its questioning of those very postulates is a clear evidence of the intellectual honesty and rationalism of the *Veda*. Coming to our discussion on *Aruṉapraṣna*, its creation story is centered around *Prajāpati*.

Before we delve into the cosmology of *Aruṉapraṣna*, we need to understand two concepts. First, *Aruṉapraṣna's* chief purpose is not to explain the exposition of the origins of universe or the history of earth. Its purpose is to describe the *Aruṉaketukāgni* ritual. Everything that is discussed therein is one way or the other related to that ritual. This ritual, like most *Vedic* rituals, is conducted to link the microcosm with the macrocosm for the benefit of the performer. In this particular ritual, the microcosm is the setup of the ritual and the macrocosm is the greater universe. Second, *Aruṉapraṣna's* creation story is focused on the origin of earth because its purpose is to design the rituals for benefit the inhabitants living on earth. This is an important distinguishing factor from the scientific theories, which are focused on explaining the creation of the greater universe and the theories of earth's formation are only a subset of their universe formation hypotheses.

In this chapter, I am proposing three postulates on this subject matter. First, I provide the *Aruṇapraśna Mantra*s related to the formation of the earth and then I present my postulates on this topic. In *Aruṇapraśna*, the contexts for these postulates are interspersed within the *Aruṇapraśna's* creation story. However, for presentational purposes, I have broken the continuous story into sections corresponding to the three postulates.

Sūkta 10.129 of *Ṛgveda*

नासदासीन्नो सदासीत्तदानीं नासीद्रजो नो व्योमा परो यत् |

किमावरीवः कुह कस्य शर्मन्नम्भः किमासीद्गहनं गभीरम् ॥ १॥

Before this universe was there, there was neither existence nor non-existence. There existed no activity either. There was no sky or space beyond it. There was nothing that could have covered what was to come later. Where was all what we see in the universe today? In whose keeping was all this then? What cosmic waters of unfathomable depths existed then?

न मृत्युरासीदमृतं न तर्हि न रात्र्या अह्न आसीत्प्रकेतः |

आनीदवातं स्वधया तदेकं तस्माद्धान्यन्न परः किञ्चनास ॥२॥

Neither death nor immortality existed then. There was no perception of either the day or of the night. There was life creating potentiality but it was not manifested as physical breath. It had within it all the energy of the manifested world and there was nothing that was subtler than this potentiality.

तम आसीत्तमसा गूहळमग्रे प्रकेतं सलिलं सर्वाऽइदम् |

तुच्छ्येनाभ्वपिहितं यदासीत्तपसस्तन्महिनाजायतैकम् ॥३॥

Before creation there was only darkness everywhere that covered everything. There was nothing perceptible about it. That *Salilam* (सलिलम्), which was everywhere covered whatever was existing in a subtle state of potentiality at that time. From that shrouded potentiality *Salilam*, one arose by the power of *Tapasyā* (तपस्या).

कामस्तदग्रे समवर्तताधि मनसो रेतः प्रथमं यदासीत् |

सतो बन्धुमसति निरविन्दन्हृदि प्रतीष्या कवयो मनीषा ॥४॥

Prior to creation, all that existed everywhere was a desire to create. This desire is similar to the desire that stems from one's mind and was the seed of the creation. The sages reflect upon this wisdom in their hearts and obtain the knowledge of the connection between the unmanifested desire and the manifested creation.

तिरश्चीनो विततो रश्मिरेषामधः स्विदासीदुपरि स्विदासीत् |

रेतोधा आसन्महिमान आसन्त्स्वधा अवस्तात्प्रयतिः परस्तात् ॥५॥

That potentiality-desire that existed prior to creation spread across widely around, above and below like the rays of the sun. It had within it a powerful creative force, which was based on self-energized strength and was fully capable of the creation that was yet to come.

को अद्धा वेद क इह प्र वोचत्कुत आजाता कुत इयं विसृष्टिः |

अर्वाग्देवा अस्य विसर्जनेनाथा को वेद यत आबभूव ॥६॥

Who knows the truth ? Who can guide us correctly on this subject? Where from this creation manifested? What is the root cause of this creation and its complex diversity and why did this all happen? The sun moon and these worlds are secondary to that root cause. So, who knows about that which caused this universe to appear everywhere?

इयं विसृष्टिर्यत आबभूव यदि वा दधे यदि वा न |

यो अस्याध्यक्षः परमे व्योमन्त्सो अङ्ग वेद यदि वा न वेद ॥७॥

What is the root cause from which this creation with all its diversity is produced and who is supporting or not supporting this manifested creation? The ruler of this creation who is established in the heavens knows about the root cause or maybe even he does not know.

Aruṇapraṣna Ḇūmī Mantras (मन्त्र)

Prapāṭaka 1 *Anuvākā* 23 *Mantra* 1

आपो॒ वा इ॒दमा॑सन्त्सलि॒लमे॒व ।

स प्र॒जाप॑ति॒रेकः॑ पुष्करप॒र्णे सम॑भवत् ।

तस्या॒न्त॒र्मन॑सि काम॒स्सम॑वर्तत ।

इ॒दँ सृ॑जेय॒मिति॑ ।

तस्मा॒द्यत्पुरु॑षो॒ मन॑साऽभि॒गच्छ॑ति ।

तद्वा॒चा व॑दति ।

तत्कर्म॑णा करोति ।

तदे॒षाऽभ्यनू॑क्ता ।

Before the world that we see around was created, there existed *Salilam* only and not Gods, humans or any other *Paṉcaḇūtā:* (पञ्चभूताः) material forms. At that time, *Prajāpati* manifested on a lotus leaf and the desire to create the world entered his mind. Therefore, human first thinks in his/her mind, speaks what he/she thinks and then uses his/her body to do that which he/she thought and spoke about. This fact is discussed in another *Ṣāḵā* (शाखा) of the Veda.

Prapāṭaka 1 *Anuvākā* 23 *Mantra* 2

काम॒स्तदग्रे॒ सम॑वर्त॒ताधि॑ ।

मन॑सो॒ रेतः॑ प्रथ॒मं यदासी॑त् ॥

स॒तो बन्धु॒मस॑ति॒ निर॑विन्दन् ।

हृ॒दि प्र॒तीष्या॑ क॒वयो॑ मनी॒षेति॑ ।

Before the creation, a strong and clear desire to create originated in

Prajāpati's mind. The unmanifested creative blueprint from previous cycle of creation manifested as *Prajāpati*'s desire for creating in this cycle of creation. Sages with control of their minds were able to understand the connection between the unmanifested universe in form of *Prajāpati*'s desire and the manifested material universe. Not just this creation but all the other creations also originated from *Prajāpati*'s desire.

Prapāṯaka 1 *Anuvākā* 23 *Mantra* 3

उपै॑नु॒न्तदुप॑नमति ।

यत्का॑मो॒ भव॑ति ।

य ए॒वं वेद॑ ।

Those who understand the importance of desire they obtain their object of desire.

Prapāṯaka 1 *Anuvākā* 23 *Mantra* 4

स तपो॑ऽतप्यत ।

स तप॑स्त॒प्त्वा ।

शरी॑रमधू॒नुत ।

तस्य॒ यन्मा॒ँसमासी॑त् ।

ततो॑ऽरु॒णाः के॒तवो॒ वात॑रश॒ना ऋष॑य॒ उद॑तिष्ठन्

ये नखा॑ः ।

ते वै॑खान॒साः ।

ये वाला॑ः ।

ते वा॑लखि॒ल्याः ।

यो रस॑ः ।

सो॑ऽपाम् ।

Having desired to create the universe, *Prajāpati* did a *Tapasyā*. Here *Tapasyā* does not mean fasting etcetera but it means he contemplated on all the details related to this creation effort. *Prajāpati* then caused his body to tremble. From his trembling body originated three families of *Ṛṣis*, namely *Aruṇā*: (अरुणाः), *Ketavā*: (केतवाः) and *Vātarāṣanā*: (वातराशनाः). From *Prajāpati*'s contemplation and according to his contemplation objects were manifested. From his nails emerged *Vekānasā*: (वैखानसाः) family of *Ṛṣis*. From his bodily hair came out *Vālakilyā*: (वालखिल्याः) family of *Ṛṣis*. From his bodily fluids such as blood etcetera emerged a tortoise.

Prapāṭaka 1 *Anuvākā* 23 *Mantra* 5

अ॒न्त॒र॒तः कू॒र्मं भू॒तँसर्प॑न्तम् ।

तम॑ब्रवीत् ।

मम॒ वैत्वङ्मा॒ँसा ।

सम॑भूत् नेत्य॑ब्रवीत् ।

पूर्व॑मे॒वाहमि॒हास॒मिति॑ ।

तत्पुरु॑षस्य पुरुष॒त्वम् ।

स स॒हस्र॑शीर्षा॒ पुरु॑षः ।

स॒ह॒स्रा॒क्षस्स॒हस्र॑पात् ।

भू॒त्वोद॑तिष्ठत् ।

तम॑ब्रवीत् ।

त्वं वै पूर्वँ॑ सम॑भूः ।

त्वमि॒दं पूर्वः॑ कुरु॒ष्वेति॑ ।

Prajāpati said to the tortoise, which was swimming about in the water, you have been born from my skin and flesh. Then the tortoise said, no only my tortoise form emerged from you but I, being always omniscient have been here before. Having said this, the tortoise showed its real

Virāta (विराट) form with infinite heads and feet. Seeing this *Virāta* form, *Prajāpati* said to him, you have been here in the past, please create the universe as you have done in the past.

Prapāṯaka 1 *Anuvākā* 23 *Mantra* 6

स इ॒त आ॒दायाप॑ः

अ॒ञ्ज॒लिना॑ पु॒रस्ता॑दु॒पाद॑धात् ।

ए॒वाह्ये॒वेति॑ ।

तत॑ आदि॒त्य उद॑तिष्ठत् ।

सा प्राची॒ दिक् ।

The tortoise took a palmful of the water that was there before the creation and put that water as a brick in the east with the *Mantra*, so be it. The *Āditya Devatā* (आदित्य देवता) emerged from that direction and that became the east direction

Prapāṯaka 1 *Anuvākā* 23 *Mantra* 7

अथा॑रु॒णः के॒तुर्द॑क्षिण॒त उ॒पाद॑धात् ।

ए॒वाह्य॑ग्न॒ इति॑ ।

त॒तो वा अ॒ग्निरुद॑तिष्ठत् ।

सा द॑क्षि॒णा दिक् ।

अथा॑रु॒णः के॒तुः प॒श्चादु॒पाद॑धात् ।

ए॒वाहि वा॒यो इति॑

ततो॑ वा॒युरुद॑तिष्ठत् ।

सा प्र॒तीची॒ दिक् ।

अथा॑रु॒णः के॒तुरु॑त्तर॒त उ॒पाद॑धात् ।

ए॒वाहीन्द्रेति॑ ।

त॒तो वा इन्द्र॒ उद॑तिष्ठत् ।

सोदी॑ची॒ दिक् ।

अथा॑रु॒णः के॒तुर्मध्य॑ उ॒पाद॑धात् ।

ए॒वाहि पू॒षन्निति॑ ।

त॒तो वै पू॒षोद॑तिष्ठत् ।

सेयन्दिक्

अथा॑रु॒णः के॒तुरु॒परि॑ष्टादु॒पाद॑धात् ।

ए॒वाहि दे॒वा इति॑ ।

ततो॑ देवमनु॒ष्याः पि॒तर॑ः ।

ग॒न्ध॒र्वा॒प्स॒रस॒श्चोद॑तिष्ठन् ।

सोर्ध्वा दिक् ।

Aruṇaketukā (अरुणकेतुका) then said, "so be *Agni* (अग्नि)" and placed water on the south direction. *Agni Devatā* emerged from that direction and that direction became the South. *Aruṇaketukā* then said, "so be *Vāyu* (वायु)" and placed water on the west direction. *Vāyu Devatā* emerged from that direction and that direction became the west. *Aruṇaketukā* then said, "so be *Indra* (इन्द्र)" and placed water on the north direction. *Indra Devatā* emerged from that direction and that direction became the North. *Aruṇaketukā* then said, "so be *Pūṣan* (पूशन्)" and placed water in the middle. *Pūṣan Devatā* emerged from the middle and that became the downward direction. *Aruṇaketukā* then said, "so be *Devas* (देवाः)" and placed water in the upward direction. All *Devas*, Humans, *Pitaras* (पितराः), *Apsarasas* (अप्सरसः), and *Gandarvas* (गंधर्वाः) emerged from the top and that became the upward direction.

Prapāṭaka 1 *Anuvākā* 23 *Mantra* 8

या वि॒प्रुषो॑ वि॒परा॑पतन् ।

ताभ्योऽसुरा रक्षाँसि पिशाचाश्चोदतिष्ठन् ।

तस्मात्ते पराभवन् ।

विप्रुड्भ्यो हि ते समभवन् ।

While filling palmful of water and placing, whatever drops of water fell down, those drops became *Asurās* (असुरा:), *Rakshasas* (राक्षसा:) and *Piṣācas* (पिशाचा:). So, they became degraded. Since they were born off the water that fell, they became fallen.

Prapāṭaka 1 *Anuvākā* 23 *Mantra* 9

तदेषाऽभ्यनूक्ता

For this topic, this *Mantra* is there in another *Ṣākā*.

Prapāṭaka 1 *Anuvākā* 23 *Mantra* 10

आपो ह यद्बृहतीर्गर्भमायन् ।

दक्षन्दधाना जनयन्तीस्स्वयंभुम् ।

तत इमेऽध्यसृज्यन्त सर्गाः ।

अद्भ्यो वा इदँसमभूत् ।

तस्मादिदँसर्वं ब्रह्म स्वयंभ्विति ।

The great water having the ancient womb and a desire to create the tortoise shaped *Paramātmā* (परमात्मा) obtained world within its womb. It is from those waters this universe was created. So the whole universe was created from these waters only. That is why the whole world is self-created, *Svaya˙ Siddha*: (स्वयं सिद्ध:).

Prapāṭaka 1 *Anuvākā* 23 *Mantra* 11

तस्मादिदँ सर्वं शिथिलमिवाद्ध्रुवमिवाभवत् ।

Water is formless and flowing and that is why the universe that was formed from these waters was not rigid and was ever changing.

Prapāṯaka 1 *Anuvākā* 23 *Mantra* 12

प्र॒जाप॑ति॒र्वाव् तत् ।

आ॒त्मना॒त्मानं॑ वि॒धाय॑ ।

तदे॒वानुप्रावि॑शत् ।

To give it certain stability *Prajāpati* with his own capability changed himself into the form of the universe and entered into it.

Prapāṯaka 1 *Anuvākā* 23 *Mantra* 13

तदे॒षाऽभ्यनू᳚क्ता

For this topic, this *Mantra* is there in another *Ṣāḵā*.

Prapāṯaka 1 *Anuvākā* 23 *Mantra* 14

वि॒धाय॑ लो॒कान्वि॒धाय॑ भू॒तानि॑ ।

वि॒धाय॒ सर्वाः᳚ प्र॒दिशो॒ दिश॑श्च ।

प्र॒जाप॑तिः प्रथम॒जा ऋ॒तस्य॑ ।

आ॒त्मना॒त्मान॑म॒भिसंवि॑वे॒शेति॑ ।

The first child of *Ṛtam*, *Prajāpati*, having created the worlds, the life forms, the directions (East Etcetera) and the cross directions (Northeast Etcetera) with his consciousness and with his material form perfectly entered into the universe he created.

Prapāṯaka 1 *Anuvākā* 23 *Mantra* 15

सर्व॑मे॒वेदमा॒प्त्वा ।

सर्व॑मव॒रुद्ध्य॑ ।

तदे॒वानु॒प्रवि॑शति ।

य ए॒वं वेद॑

One who understands this creation process as described here, that person obtains all the desires, attracts everyone and enters the universe himself.

Ḇūmī Postulates

Ḇūmī Postulate 2.1 - The entire universe self-manifested out of eternal primordial waters, *Salilam*.

Ḇūmī Postulate 2.2 - Creation and Dissolution is an infinite cycle stretching both into the past and into the future.

Ḇūmī Postulate 2.3 - *Aruṉapraṣna* 1.23.7 is a symbolic description of earth's change in axis from its North-South orientation to the present tilted orientation.

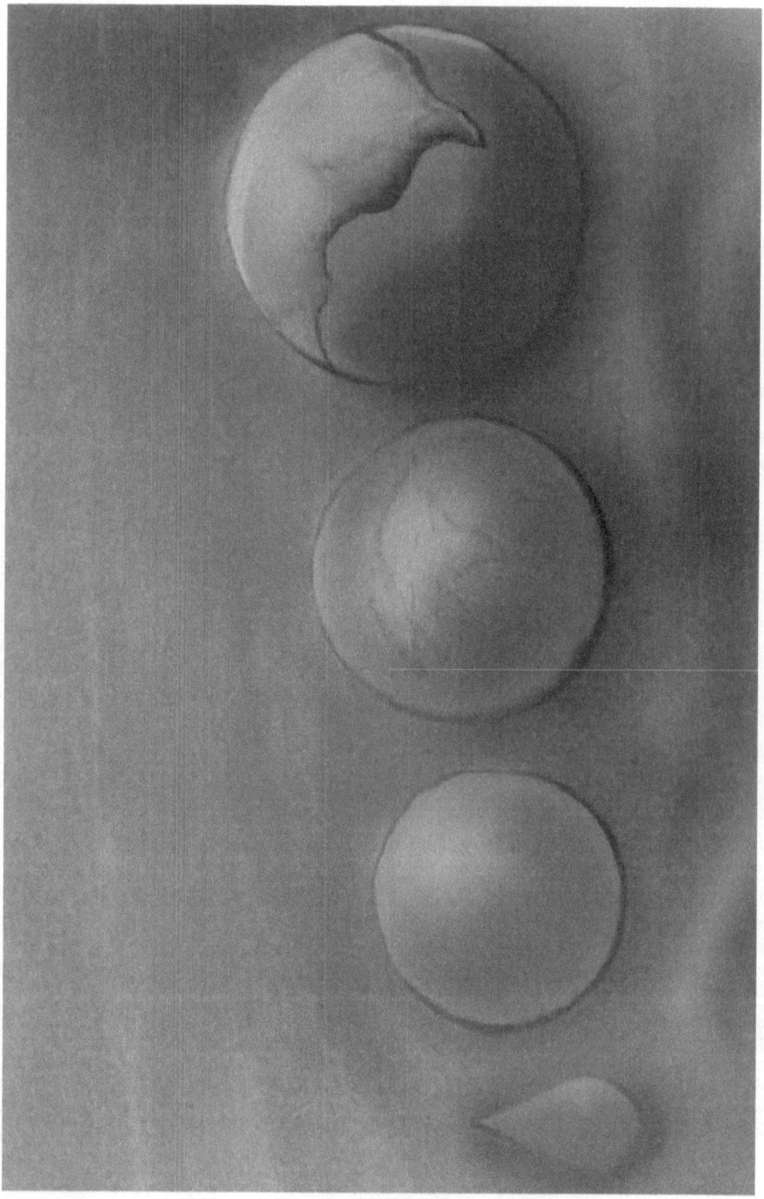

Aruṇapraṣna's Cosmology of the Earth. *Āpa* emerges from *Salilam* and the physcial Earth comes out of *Āpa*.

Bumi Postulate 2.1

Everything manifested out of *Salilam.*

Arunaprasna's creation story starts with the first *Mantra* of the 23rd *Anuvaka* with a reference to *Salilam.* Here, *Arunaprasna* states that *Salilam* is the source and cause of all the manifested universe. We can understand this *Salilam* as primordial waters. So, before there was anything, there was only these primordial waters.

According to *Vedic* cosmology, the material world is variegated combinations of the five basic substances: the *Pancabutas*, namely *Akasa* (आकाश), *Vayu*, *Agni*, *Apa:* (आप:), and *Prtvi* (पृथ्वी). *Akasa* the container space element, *Vayu* the air element, *Agni* the fire element, *Apah* the water element, and *Prtvi* signifies the earth element. According to *Arunaprasna*, even these elemental *Pancabutas* were dormant within *Salilam*. So, what exactly is this primordial substance *Salilam*? *Arunaprasna* does not provide any further insight into this subject.

Sanskrit words have contextual meanings and the one that makes the most sense here defines *Salilam* as something that is active and flowing. But we should not make the common mistake of equating it with water or any other familiar substance. Existing scientific knowledge does not have an equivalent description of *Salilam*. It is neither the interstellar gas and dust out of which all the stars, planets and other objects of the universe are thought to be formed, nor is it the building blocks of the elements, Quarks and Gluons. It is an undiscovered core substance from which all of what we see today manifested including time.

The interrelated scientific concepts such as quantum bridge, loop quantum gravity, quantum geometry and big bounce seem to be heading in that direction and may pave the way for science to understand *Salilam* one day.

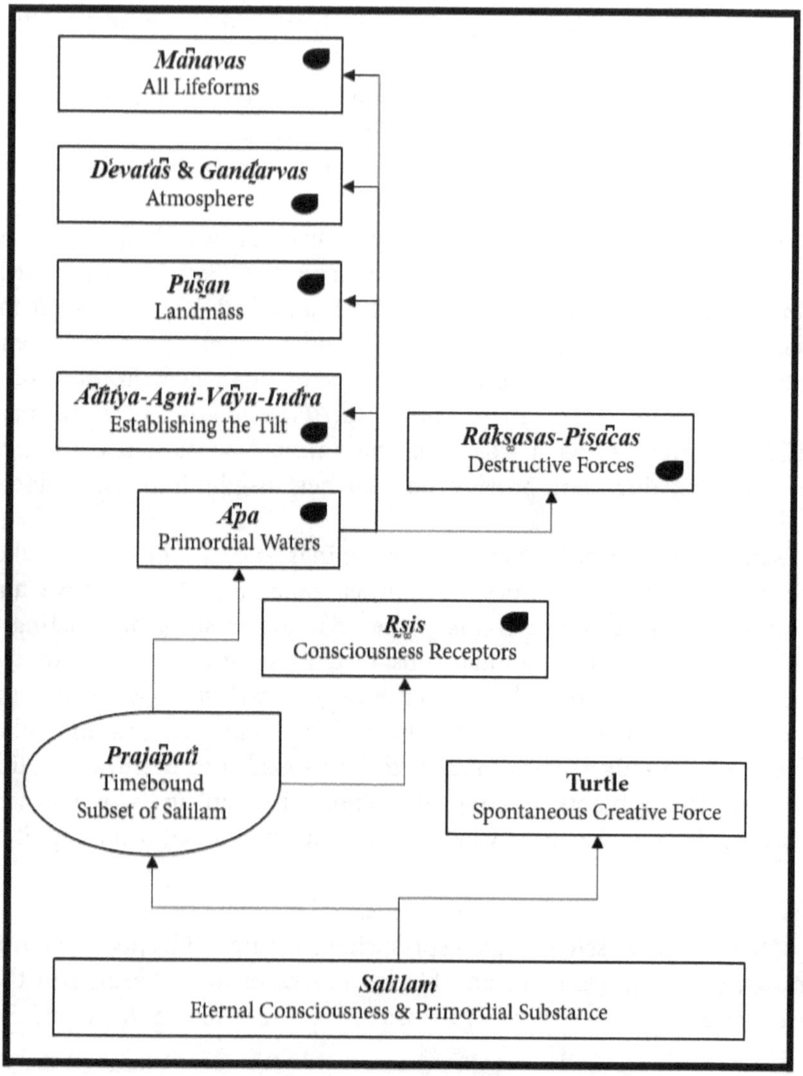

Aruṇapraṣna's Cosmology of the Earth from *Salilam* to Lifeforms.

Būmī Postulate 2.2
Creation is cyclical

In the first postulate we learnt that everything manifested from the primordial waters, *Salilam*. In this second postulate, we continue our creation story starting with a discussion of a *Vedic Devatā, Prajāpati*. Veda does not have any one supreme *Devatā*, but it does describe *Prajāpati* as a creator of the universe, particularly in the *Ṛgveda Sūkta* 10.121. In *Aruṇapraṣna*, it is also *Prajāpati*, who first emerges on a lotus leaf from *Salilam*. Upon emerging, a desire to create takes birth in *Prajāpati*'s mind. This awakening of his creative desire jump starts the creation cycle. Here we touch on *Aruṇapraṣna*'s metaphysics of creation. Despite significant advances in Cognitive Sciences and Quantum Physics, there is a substantial gap that science must bridge before we can attempt to understand *Prajāpati*'s creative desire in scientific terms. This desire can be loosely understood as an instruction set for creation. This set is not entirely new but is a permutation and combination of the instruction sets of all prior creations and dissolutions, which lie embedded within the *Salilam*. As per Hindu Cosmology, this observable universe is like a time bound wave on the immense ocean of *Salilam*, from which creation waves have been rising and falling eternally and will eternally continue to do so in the future.

In other words, *Prajāpati* embodies a subset of the eternal consciousness, *Salilam*. One such subset was the seed of the creation wave of our current observable universe. The lotus leaf on which the *Prajāpati* emerges from the *Salilam* poetically illustrates this subset concept. It is important to understand that *Prajāpati* and all other animate and inanimate descriptions in this story as anthropomorphic overlays on the cosmological understanding of the *Vedic Ṛṣis*.

Having thus developed a desire to create the world, *Prajāpati* enters a state of *Tapasyā*. *Sayanacharya* in his *Bāṣya* describes this *Tapasyā* as a process of *Prajāpati* contemplating on the details of his creation. After a long time in the *Tapasyā*, *Prajāpati*'s body starts to vibrate. From these vibrations, five different families of *Ṛṣis* emerge from *Prajāpati*'s body. So, the subset of consciousness having emerged from *Salilam*, undergoes a long period of gestation during which some yet to be understood processes occur. These processes create the blue print of the universe to be formed from the consciousness subset. Here, *Aruṇaprasṉa* through the concept of *Ṛṣis* maybe describing the transformation of the subtle consciousness into physical forces and energy fields that laid the foundation of the physical universe. We can also conceptualize the *Ṛṣis* as the physical receptors of the consciousness subset. So, the creative consciousness acts through these physical forces and energy fields. More research is needed to understand this part of the creation story in *Aruṇaprasṉa.*

Prajāpati's body not only gives birth to the *Ṛṣis* but also water, which is the crucial ingredient for the creation recipe of both science and *Aruṇaprasṉa.* This is an important detail within *Aruṇaprasṉa's* cosmology, which explains that water forms the basic building blocks of our observable universe, and that this water was created directly from the consciousness subset that emanated from *Salilam.* This emanation process is metaphorically indicated as happening through the bodily fluids of *Prajāpati, Rasā:* (रसाः). This contrasts with the baryogenesis model of the progressive development of oxygen from lighter elements, which then combines with hydrogen to form water via fusion. *Aruṇaprasṉa's* 22nd *Anuvākā* describes the presence of water in every aspect of the observable universe. It must be pointed out that these waters are not the same as the water we know today. Even when we scientifically analyze the various types of waters we know today, we learn that all waters are not the same. Scientists have sampled waters from deep inside the earth and have found that the percentage of deuterium oxide is much higher than in surface water. Scientists have discovered several other states of water beyond the commonly known three states. Some scientists believe that more water exists in the earth's mantle than in the oceans. So, these waters that originated out of *Salilam* have different characteristics perhaps at the quantum level than the waters we are familiar with. That water's characteristics were such that they made it the life-giving raw material for the formation of the earth. *Aruṇaprasṉa's* creation story describes the development of earth from these waters.

Prajāpati then notices a turtle floating around in the waters and assumes it emerged from his body as well. They then have a conversation where the turtle tells *Prajāpati* that only its material form emerged from his body. It declares itself to be eternal and the embodied form of all creation and demonstrates its claim by giving *Prajāpati* a glimpse of the entire universe within itself. *Prajāpati* then accepts the turtle's eternal and omnipotent nature and requests it to recreate the universe. Here, *Aruṇapraṣna* uses the analogy of a turtle to show the spontaneity of the creation process. The turtle can withdraw its limbs and head into its shell and lie motionless for extended periods of time. In this state, it can be mistaken for an inanimate object but that illusion is dispelled when it starts to move. Through this turtle, *Aruṇapraṣna* reiterates that this creation cycle did not create anything that did not exist in earlier creation cycles. It purposefully breaks the cause-and-effect sequence to illustrate the limitation of human understanding of the creation process. The unpacking of the physical universe from the subtle consciousness happens spontaneously without the need for a creating force. Having grasped the truth of the turtle's theory, *Prajāpati* asks him to manifest creation.

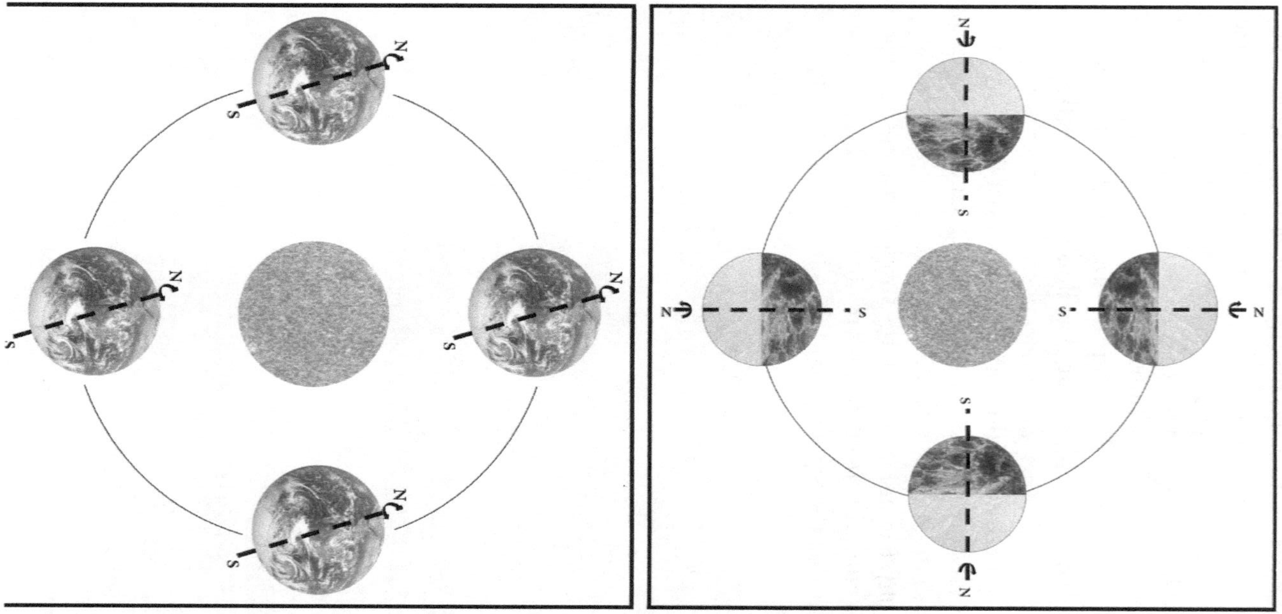

Earth's axis of rotation changes from North - South to the present slanted Northeast -Southwest.

Image developed using: (1) Glowing `a`a lava flow front advancing over pahoehoe lava on the coastal plain of Kilauea Volcano, Hawaii, USGS[2], Public domain, via Wikimedia Commons. (2) A photograph of the snow surface at Dome C Station, Antarctica, Stephen Hudson, CC BY-SA 3.0 <http://creativecommons.org/licenses/by-sa/3.0/>, via Wikimedia Commons (3) Sun, NASA/SDO (AIA), Public domain, via Wikimedia Commons.

B̰ūmī Postulate 2.3

Earth's axis changed from North-South to the tilted axis the earth has today.

The turtle starts the seven-step creation process by first taking a handful of water and offering them in the easterly direction, causing the *Āditya Ḋevatā* to rise and designate the east direction. It then offers the waters to the south and brings forth the *Agni Ḋevatā* and the south direction. Similarly, the west with the *Vāyu Ḋevatā*, and the North with the *Indra Ḋevatā* follow. Thereafter, the turtle offers the waters in the downward direction and brings forth the *Pūṣan Ḋevatā*. Finally, it offers waters in the upward direction and manifests all life forms including the *Ḋevatā*s, *Pitaras*, *Gaṇḍarvas*, *Apsarās* and humans. During this effort, some drops of water inadvertently fall out its offering hands and these drops manifest as the *Asurās*, *Rākṣasās* and *Pisācās*. Thus, the entire universe self-manifested out of the waters. Because of its formation from water the universe lacked stability and flowed like water. *Prajāpati* then takes the shape of the manifested universe and merges with it, giving it the needed stability.

How do we interpret *Aruṇapraṣna*'s creation story vis-à-vis what cosmology tells us? The scarcity of detail in *Aruṇapraṣna* gives scope for multiple explanations. I am providing one possible model that attempts to explain the turtle's creation process in scientific terms. This is not an attempt to reconcile *Aruṇapraṣna* with science. It is a distinctly different model than one based on the Big Bang, baryogenesis and nebular hypothesis theories. In our scientific model of *Aruṇapraṣna*, earth starts as a spherical body of water revolving around the sun. We interpret the turtle manifesting the sun and the east direction as some event that created the tilt of the earth's axis of rotation. Prior to this event, earth's axis, if it existed, was such that the North Pole always faced away from the sun and therefore was frozen, and the South Pole was always exposed to the sun and so existed as water vapor. We are basing this assumption on the turtle manifesting *Indra Ḋevatā* from the North and *Agni Ḋevatā* from the South. *Agni* here signifies energy and heat. Since only the South was exposed to the Sun all along, earth's energy source was in the South. The tilt and subsequent increased mixing of hot and cold waters combined with east to west rotation may have created waves of water vapor that is described as *Vāyu* from the west.

The turtle then manifests the *Pūṣan Ḋevatā*, which we are interpreting as emergence of land mass from within the body of water. Substantial research is needed to understand how this can actually happen. This emergence of solid land from water is a perspective unique to *Aruṇapraṣna*. We can take a cue from the process of photosynthesis where gaseous carbon dioxide, liquid water and sunlight produce solid carbon material. Similarly, there must be some unknown process through which solid land mass is eventually manifested out of these first waters that constituted earth. Maybe a different form of baryogenesis and a photosynthesis like process occurred within large masses of water to manifest various elements that combined to form the land mass. Subsequent to formation, the turtle asks the *Ḋevatas*, *Pitaras*, *Ganḍarvas*, *Apsarasas*, and humans to emerge from the top. We interpret this as the formation of the atmosphere composed of the various layers depicted as the *Ḋevatas*, *Pitaras* and *Ganḍarvas*, as well as the emergence of life forms. This atmospheric model is presented in the *Vāyumaṇḍalam* chapter. The final step, where *Prajāpati* enters all creation to give it a form, is interpreted as a sum of all the various permutations and combinations that occurred in the created universe to establish the necessary dynamic balance in it.

Conclusion

The universe we know and the universe we have yet to discover is all self-manifested out of the eternal primordial waters, namely *Salilam*. Creation is an infinitely cyclical process that has been and will continue to occur for eternity. *Aruṇapraśna* 1.23.7 is a symbolic description of the change in axis of the earth from North-South orientation to the present-day tilted orientation.

The potential areas for further research based on this topic include:

- Comprehensive study of characteristics of *Salilam* in the *Vedic* and *Purāṇic* corpus.
- Comparative study of *Salilam* in the *Veda* and *Purāṇas* and Cosmic Waters in the various creation stories from around the world.
- Comparative study of *Prajāpati* in *Aruṇapraśna's* creation model and the phenomena of big bounce after contraction and beginning of next expansion in other cyclical cosmological models.
- What are the possible cosmic events that could have impacted earth's tilt? Are these events internal or external or a combination?
- Exploration of frameworks for scientific understanding of the development of landmass from the water that came out of *Prajāpati* in *Aruṇapraśna*.

Chapter 3
Vāyumaṇḍalam

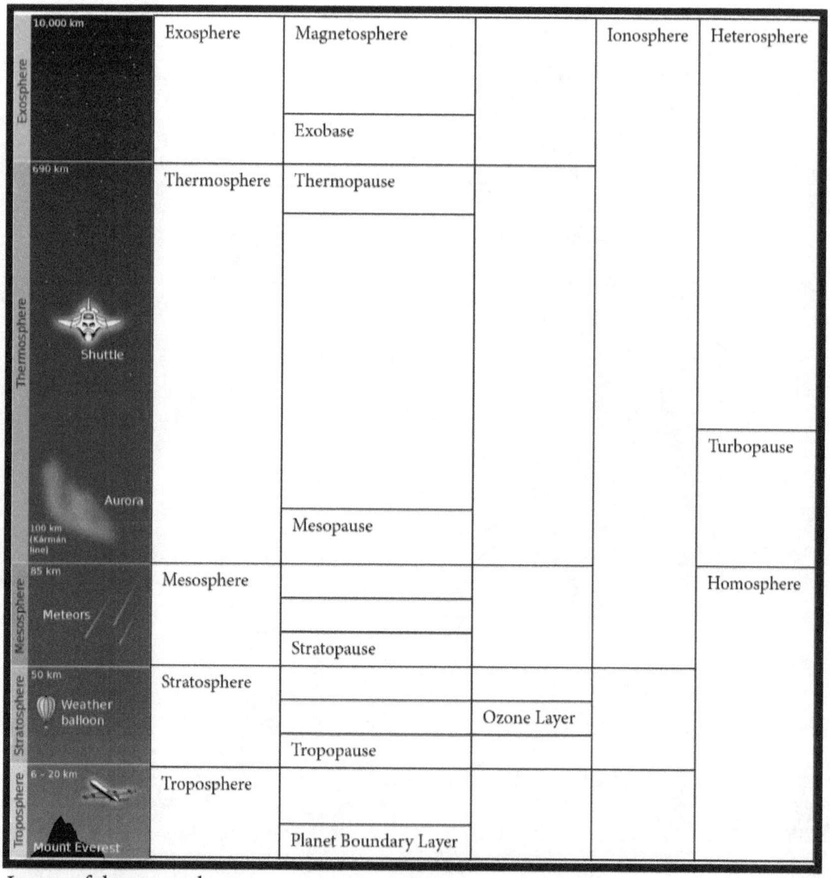

Layers of the atmosphere.

NOAA & User:Mysid, Public domain, via Wikimedia Commons.

Science

The Earth's atmosphere is a spherical, gaseous envelope that covers its surface and ten thousand kilometers of space above it. It protects the life on Earth in multiple ways including functioning as a pressure blanket that enables liquid water to exist on Earth, shielding the planet from cosmic and solar radiation, keeping it warm via the greenhouse effect, and reducing diurnal temperature variation. The atmosphere also serves as a conduit and repository for the oxygen and carbon dioxide from the cellular respiration and photosynthesis cycles respectively. To put the scale of this in perspective, somewhere between 50 to 75 percent of oxygen is generated by algae and cyanobacteria in the oceans and the rest comes from terrestrial plants. The influence of winds on global weather patterns including the ocean currents and water cycle cannot be understated. Apart from physics and chemistry with a principal focus on weather forecasting, the study of atmosphere includes climatology and aeronomy.

Atmospheric scientists have divided the atmosphere into 5 primary layers and several overlapping secondary layers based on chemical composition, density, pressure, temperature, and other variables. These layers have loosely defined boundaries, and the demarcations vary with latitude and the seasons. The first layer, Troposphere is the layer hugging the earth's surface. Containing about 75% of total mass of the atmosphere and 99% of the water vapor, it is the theatre for all the weather phenomena. The planetary boundary layer, which is the bottom of the Troposphere is influenced by frictional interaction with the earth's surface while the top of the Troposphere houses the Tropopause, which is a temperature inversion layer where atmospheric temperature remains static with change in elevation. Above the Tropopause is the Stratosphere, where the temperature rises with increase in elevation. Aircraft use this layer to avoid the weather turbulences in the troposphere. The ozone layer, about which we will discuss in greater detail later is a part of the Stratosphere. The next layer is the Mesosphere. Temperature falls with elevation in this region. East-West winds, atmospheric gravity waves and planetary waves in this layer drive the global air circulation on our planet. The region includes various metal layers including sodium, potassium, and iron. These are byproducts of metallic sublimations from thousands of ablating meteors. The meteoric smoke from the ablation also creates condensation nuclei for noctilucent clouds. Beyond the Mesosphere, comes the Thermosphere. It has extremely low density and the residual atmospheric gases in this layer sort into strata based on their molecular mass. Highly energetic solar radiation in Thermosphere causes thermospheric temperatures of the gases to increase with altitude but the low density prevents the conductive transfer of the same. Intense solar radiation here causes photoionization of molecules. The outermost layer is the Exosphere. Here the earth's atmosphere thins out and merges with outer space. It is mostly composed of hydrogen and helium. These layers are not clearly demarcated regions, but have transition regions between them. In addition, scientists superimpose secondary classifications on the primary classification into the above described five layers. These include Ozone Layer, Ionosphere, Magnetosphere, Heterosphere, and Homosphere.

Aruṇapraṣna

The *Ḇūmī* chapter discusses the formation of atmospheric layers as a part of the Earth creation story where *Prajāpati*, through the *Aruṇaketukā* in the form of a tortoise, establishes the *Devatās* in the upward direction. This can be understood as a symbolic association between the atmospheric layers and the *Devatās*. More specifically, *Aruṇapraṣna* 1.5.3 to 1.5.5 conceptualizes the loss of balance and the subsequent rebalancing of the ozone-oxygen cycle through a symbolic story of severing and reattachment of *Rudra's* head. In this chapter, I am proposing three postulates on this subject matter. They are outlined below the related *Aruṇapraṣna Mantras*.

Aruṇapraṣna Vāyumaṇḍalam Mantras

Prapāṭaka 1 *Anuvākā* 5 *Mantra* 3

दि॒व्यस्यैक॒ा धनु॑रार्त्निः ।

पृ॒थि॒व्यामप॑रा श्रि॒ता ॥

तस्येन्द्रो वम्रि॑रूपे॒ण ।

ध॒नुर्ज्यामच्छि॒नथ्स्वय॑म् ॥

One end of the bow of *Aruṇaketukāgni*, who has the form of the blazing tongue of *Viṣvedeva* (विश्वेदेव), is situated in the heavens and the other end of that bow is on the earth. Scared of the bow, *Indra* in the form of a termite eats away the bow string.

Prapāṭaka 1 *Anuvākā* 5 *Mantra* 4

तदि॑न्द्र॒धनु॑रित्य॒ज्यम् ।

अ॒भ्रवर्णे॑षु चक्ष॑ते ।

एतदेव शंयोर्बार्हस्पत्यस्य ।

एतद्रुद्रस्य धनुः ॥

That multicolored stringless bow now in the sky is called *Indra*'s bow because it was created as explained by *Indra* in the previous verse. This is the bow of *Ṣamyu* (शम्यु) the son of *Bṛhaspati* (बृहस्पति) as well. This is the bow of *Aruṇaketukāgni* who is called *Rudra*.

Prapāṭaka 1 *Anuvākā* 5 *Mantra* 5

रुद्रस्य त्वेव धनुरार्त्निः ।

शिर उत्पिपेष ।

स प्रवर्ग्योऽभवत् ।

तस्माद्यस्सप्रवर्ग्येण यज्ञेन यजते ।

रुद्रस्य स शिरः प्रतिदधाति ।

नैनँरुद्र आरुको भवति ।

य एवं वेद ॥

The strike of the unstrung bow staff severed the head of *Rudra* and flung it into the sky and when it smashed to the ground it became powdered by the impact. This powdered head became *Pravargya* (प्रवर्ग्य). Therefore, one who performs *Yajṇa* with this *Pravargya*, joins this head back to *Rudra's* body and *Rudra* does not harm that performer.

Vāyumaṇdalam Postulates

Vāyumaṇdalam Postulate 3.1 - The story of severing of *Rudra*'s head in *Aruṇapraṣna* 1.5.3 to 1.5.5 is a metaphorical representation of the ozone layer depletion.

Vāyumaṇdalam Postulate 3.2 - The *Pravargya* ritual mentioned in *Aruṇapraṣna* 1.5.5 helps in restoring the depleted ozone layer.

Vāyumaṇdalam Postulate 3.3 - I propose a new analysis model for atmospheric layers based on *Vedic Devatās*.

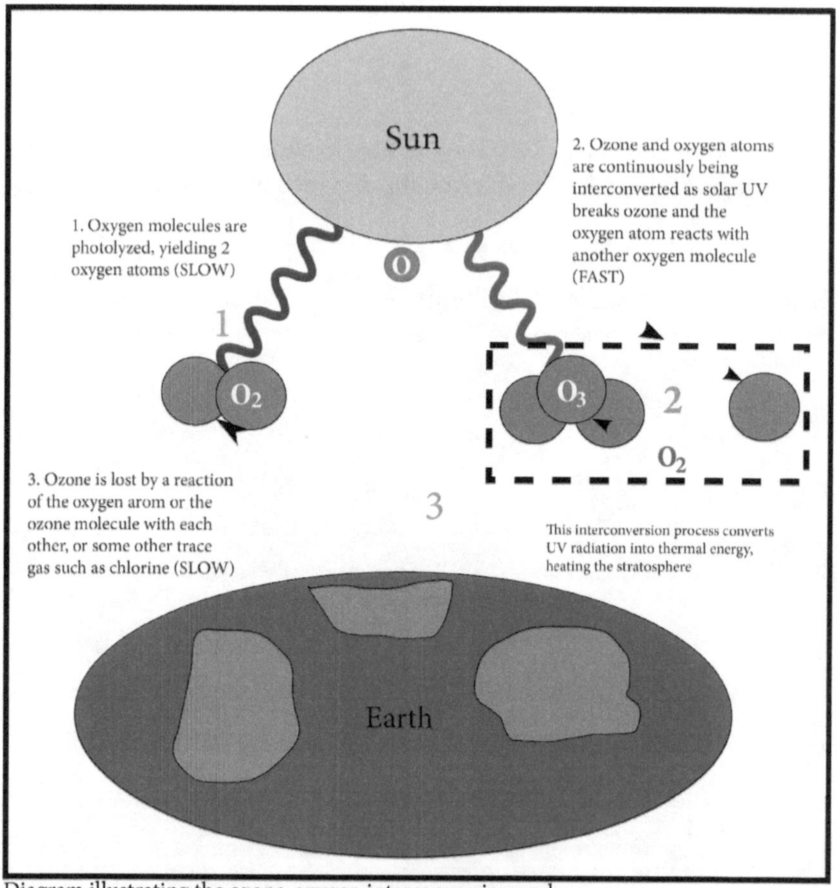

Diagram illustrating the ozone-oxygen interconversion cycle.

NASA, Public domain, via Wikimedia Commons.

Vāyumaṇḍalam Postulate 3.1

Severing of *Rudra's* head is ozone layer depletion

The Ozone Layer is in the lower stratospheric region, about 15 to 35 kilometers above the Earth's surface. It contains 90% of the ozone in the atmosphere, and the remainder can be found throughout the troposphere and stratosphere. Its thickness varies with changes in latitude, longitude and the seasons. Temporal and spatial variations are primarily driven by atmospheric circulation patterns and variations in solar UV radiation intensity. It is thinner near the equator and thicker near the poles, except near the South Pole, where it thins out to create a hole, a byproduct of human activities and various pollutants. The majority of ozone is produced over the tropics and is slowly transported to the polar regions by stratospheric winds. So, both the concentration of ozone molecules and the thickness of the ozone layer are increased in the mid and higher latitudes. Longitudinal spatial variations have somewhat more complex patterns and cannot be generalized the same way. Coming to the temporal variations, we can see that they increase as we move towards the poles. While the ozone layer over the tropics is more or less stable, the ozone layer over the mid and higher latitudes has significant changes in its thickness/density over the course of a year. However, even with seasonal changes, the thickness of the ozone layer in the mid and higher latitudes is always higher compared to the tropics.

The ozone in the atmosphere is created as a part of the ozone-oxygen interconversion cycle that occurs continuously in nature. This cycle creates oxygen from ozone and ozone from oxygen in the presence of solar UV-B or UV-C radiation. Through this continuous interconversion process, the ozone layer absorbs and converts all UV-C and most of the UV-B solar radiation to thermal energy. Both types of radiation are extremely harmful to all lifeforms that live on the Earth's surface. The overall amount of ozone in the stratosphere is determined by the balance between production of ozone by solar radiation and its removal from the atmosphere, as outlined above. Under most natural circumstances, the ozone removal rate is slow since the concentration of oxygen atoms is very low. This maintains the balance within the ozone-oxygen interconversion cycle. However, certain free radicals, primarily molecular radicals of hydroxyl (OH) and nitric oxide (NO) and atoms of chlorine (Cl) and bromine (Br) disrupt the ozone-oxygen interconversion cycle by increasing the ozone removal rate and causing ozone depletion. Hydroxyl and nitric oxide radicals are naturally occurring. Chlorine and bromine atoms can be produced from volcanic eruptions, ocean water interactions with the atmosphere and other known and unknown natural processes. However, the current alarming concentrations of chlorine and bromine are from manmade chlorofluorocarbons and halons. Each chlorine or bromine atom can catalyze tens of thousands of ozone decomposition reactions before it is removed from the stratosphere.

Here is how *Aruṇapraṣna*'s story behind this postulate goes. *Rudra* is a *Vedic Devatā* associated with storms, wind, war, hunting and healing, amongst other things. As per *Aruṇapraṣna*, *Rudra* and *Bṛhaspati* are both the sons of *Būmi*, the Earth, and *Dyośpitṛ* (द्यौष्पितृ), the Sky. Elsewhere in the *Veda*, the *Devatās* of storms, the *Marutas*, are considered *Rudra*'s sons. Once upon a time, after an intense battle, *Rudra* was resting on one end of his massive bow. The end on which *Rudra* was resting his head was in the sky and bow's other end was on the earth. *Indra*, the king of *Devatās*, assumes the form of an antlike insect and chews out the bow string. The snapped string causes the unstrung bow staff to straighten out and sever *Rudra*'s head, which rises up and then falls down to the Earth completely powdered. The unstrung bow then became the rainbow we see in the sky, and the powdered head became the oblation for the *Pravragya*, an ancillary ritual within the *Somayāga Ṣrota* ritual. When this *Pravragya* ritual is performed, *Rudra*'s head is reattached to his body. Those who perform this ritual will please *Rudra*, and will not be susceptible to his wrath. *Sāyaṇācārya*, in his *Bāṣya* on *Aruṇapraṣna*, interprets *Rudra* to be a synonym for *Aruṇaketukāgni*, the main subject matter of *Aruṇapraṣna*. The 4th and 5th *Prapāṭakas* of the *Tettirīya Āraṇyakam* are dedicated to the *Pravargya* ritual. A similar story is told there with *Rudra*'s place being taken by another *Vedic Devatā*, *Viṣṇu*, who is identified as the *Yajṇapuruṣa*. In this version of the story, we are given a little bit more detail on the preceding conflict between *Viṣṇu* and the rest of the *Devatās*. While this conflict is not presented in *Aruṇapraṣna*'s version of the story with *Rudra*, *Sāyaṇācārya* elaborates, saying that *Indra* assumes the form of an antlike insect because he fears that *Rudra*'s massive bow will incite conflict amongst the gods.

Regardless of whether it is *Rudra*, *Viṣṇu* or *Aruṇaketukāgni*, the story depicts a conflict between two competing interests that creates an imbalance in nature. Ozone molecules exist all the way from Earth's surface to the ozone layer. The bow can be understood as the atmospheric path through which surface phenomena influence the ozone layer. The antlike insects symbolize the free radicals in the atmosphere that disrupt the ozone-oxygen interconversion cycle and cause ozone depletion, depicted as the severing of *Rudra*'s head. Ozone disassociation requires UV radiation, which only exists at the altitude of the ozone layer. Therefore, the antlike insects chewing the bowstring sever only the head, without affecting any other part of the body. Today, our understanding of free radicals is limited to a few compounds and molecules such as OH, NO, Cl and Br, but there may be many other yet to be discovered radicals that naturally occur in the atmosphere and cause ozone depletion. These radicals may be created by natural phenomena such as volcanoes or more importantly, for our story, by human actions such as deforestation, agriculture, and husbandry.

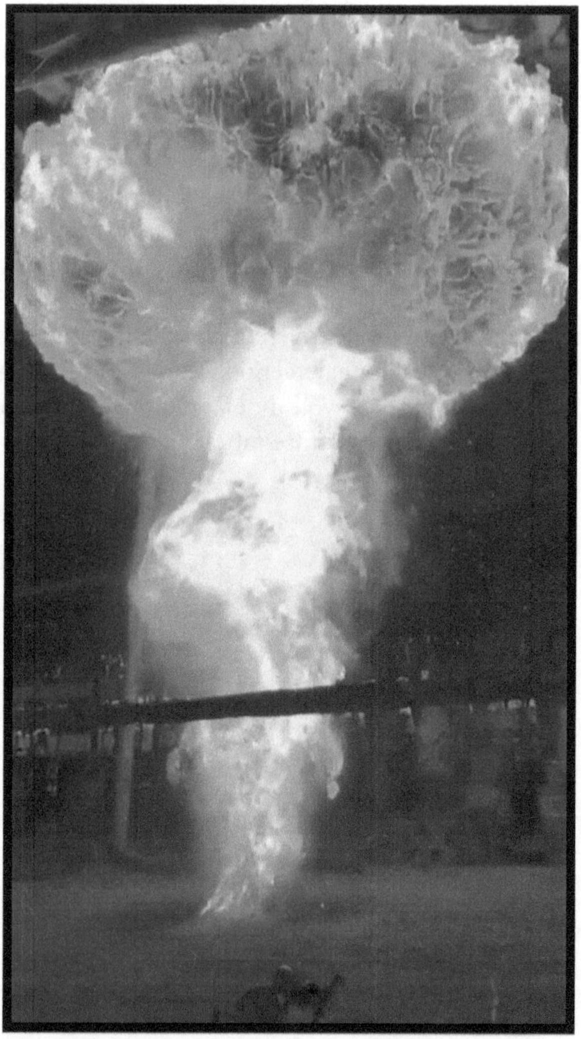

Pravargya Ritual in Peddapuram, AP, Bharat.

Photo snap from the video, Prvaragya Homam recorded by N. Mishra Studios.

Vāyumaṇdalam Postulate 3.2

Pravargya ritual restores ozone layer

In *Aruṇapraṣna* 1.5.5, the story of severing *Rudra*'s head is completed with the mention of the *Pravargya* ritual. This ritual is presented in much greater detail elsewhere in *Tettirīya Āraṇyakam* and is a major component of most *Ṣrōta Yajṇas*. *Yajṇic Karmakāṇdas* are prescribed for the *Gṛhasṭa* (गृहस्थ) by *Veda*. They help us to live in accordance with *Ṛtam* by restoring the balance in nature, after it is disrupted by human actions. One of the restoring rituals designed by Hindu ancestors is the *Pravargya* ritual. *Aruṇapraṣna* says that this ritual recreates and reattaches *Rudra*'s head to his body. The ritual requires heating a specially prepared humanoid shaped earthen pot with clarified butter, or ghee, in it until it is red hot. A prescribed quantity and mix of goat and cow milk, *Ģarma* (घर्म), is then poured into this red-hot pot. The *Ģarma* instantaneously catches on fire and a massive flame shoots several feet upwards. If we assume that *Rudra* symbolizes the ozone in the atmosphere, then we can have a scientific understanding of the *Pravargya* ritual. The ritual transmits into the atmosphere particles and/or ions that remove the free radicals, which are disrupting the Ozone - Oxygen interconversion cycle balance.

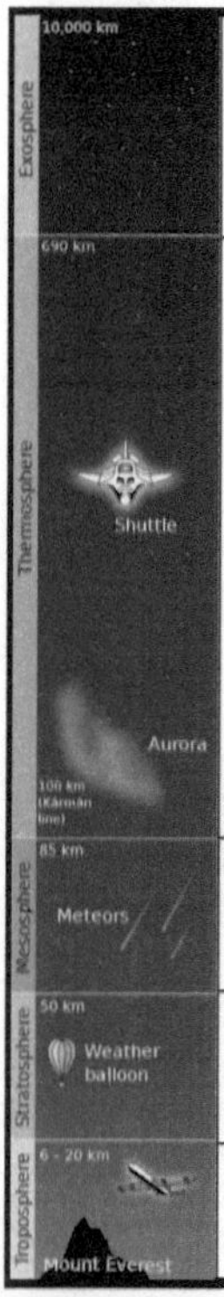

Exosphere	Magnetosphere		Ionosphere	Heterosphere	*Ganḍarvās* Layer
	Exobase				*Pitṛ Devatās* Layer
Thermosphere	Thermopause				
				Turbopause	
	Mesopause				
Mesosphere				Homosphere	*Indra* Layer
	Stratopause				
Stratosphere					*Rudra* Layer
		Ozone Layer			
	Tropopause				*Varuṇa* Layer
Troposphere					
	Planet Boundary Layer				

Proposed Atmospheric Layers based on *Veda* mapped to the Atmospheric Layers based on Science.

NOAA & User:Mysid, Public domain, via Wikimedia Commons. Last column of the table added by author.

Vāyumaṇdalam Postulate 3.3

Vedic Model of Atmospheric Layers

There are references in various places in *Aruṇaprasna*, which talk about the protection offered by various *Vedic Devatās*, which may be indicative of their symbolic associations with other atmospheric layers such as the thermosphere and magnetosphere. While a significantly more research of *Aruṇaprasna* and other *Vedic* texts is needed to assess this linkage, I am proposing an atmospheric layer analysis model that classifies the atmosphere into layers based on *Vedic Devatās*. I believe that this model will stimulate new avenues of research in the atmospheric sciences and enable a broader and deeper scientific research of the *Vedic* literary corpus.

The model classifies the atmospheric envelope into 5 spherical layers: the *Varuṇa (वरुण)* Layer, the *Rudra* Layer, the *Indra* Layer, the *Pitara Devatās* Layer and the *Gandarvās* Layer. A detailed work discussing these layers will be coming out in a separate publication following this book. The table on the left maps these *Vedic Devatā* layers to the ones as per Science.

Conclusion

The atmospheric blanket is sine qua non for life to exist on Earth. The performance and sustenance of the ozone layer shield within the atmosphere is dependent on the equilibrium maintained by the continuous oxygen-ozone interconversion cycle. Ozone depletion results from disruption in this cycle, metaphorically represented through the story of *Rudra*'s head in *Aruṇapraṣna*, and provides the *Pravargya* ritual as a solution to restore balance. Further research into this area and the development of an an atmospheric model based on *Vedic Devatās* will greatly benefit research in both the atmospheric sciences and the *Veda*

The potential areas for further research based on this topic include:

- Comprehensive study of atmospheric pollution in the *Vedic* and *Purāṇic* corpus.
- Empirical study of atmospheric effects of *Pravargya* ritual, in particular the ritual's impact on the Ozone layer.
- Mapping the characteristics of various *Vedic* and *Purāṇic Devatās* to Atmospheric Layers and Processes.

Chapter 4
Varṣā

Science

Water, being fundamental to photosynthesis, cellular respiration and cellular metabolism is essential for all known life forms. Approximately 97 percent of all water on our planet is saline and therefore not suitable for terrestrial beings dependent on freshwater. About 99 percent of the freshwater on Earth exists as glaciers, polar ice caps or ground water. 70 percent of the remaining 1 percent of freshwater is either ground ice or permafrost. A very small portion of fresh water is available as lakes, rivers, or other surface features. A miniscule amount of water also exists within various living things and of course in the atmosphere. Here is the kicker though: all fresh water on land, including subterranean water, is directly or indirectly replenished only through precipitation and rainfall constitutes more than 99 percent of it.

So, water is life for us, terrestrial beings, and rainfall is its root source. All civilizations of the past were, those in the present are, and those in the future will be dependent on rain or its variants. Then no surprise that rain is intimately connected with the rituals, cultures, and beliefs of various ethnic religions around the world. It has been a subject of human enquiry for thousands of years and continues to be an important topic for scientists today. Commercially, rainfall prediction and its measurement are a significant portion of the rapidly growing weather industry.

Water continuously circulates amongst the four spheres of the planet, the atmosphere, hydrosphere, lithosphere, and biosphere while transiting through its three states of matter: solid, liquid and gas. Rainfall is the principal part of this process and is intimately connected to all the spheres. Research on rainfall is therefore a multidisciplinary effort spanning various fields. That being said, most of the work in this arena has come out of the Atmospheric Sciences.

On a simple scale, the water cycle can be explained as a continuous loop of three processes: evaporation-transpiration-sublimation, condensation, and precipitation. Heat from the sun causes water to evaporate from

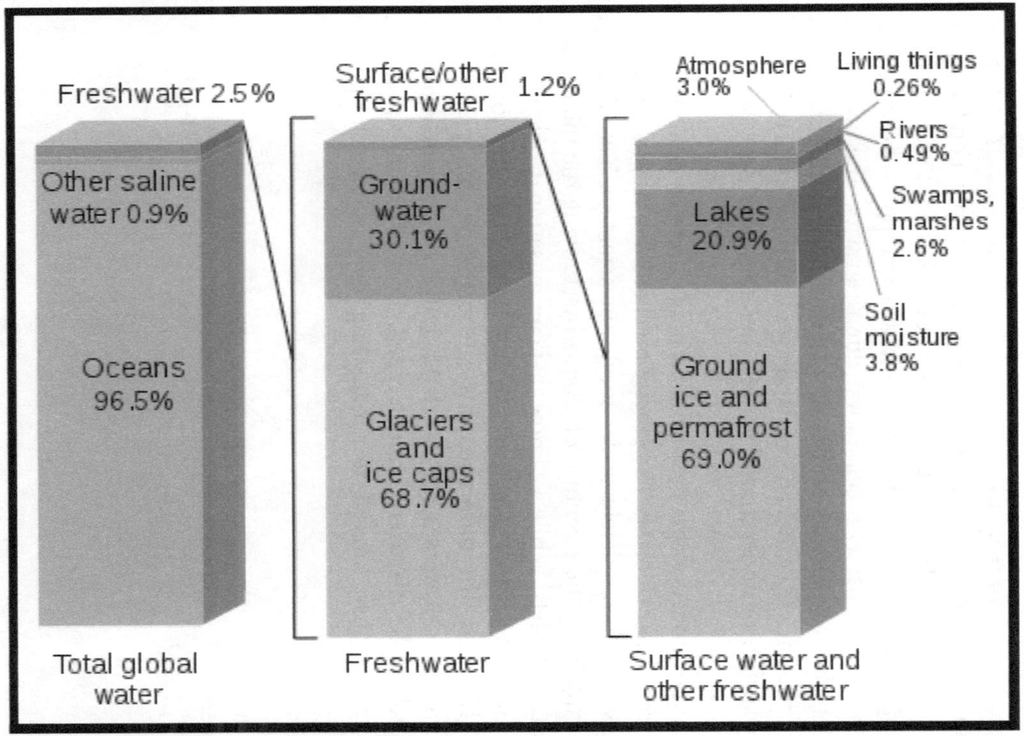

USGS, Graph of the locations of water on Earth.

USGS, Public domain, via Wikimedia Commons.

various bodies of water, principally from the oceans, and wet soil into the atmosphere. Some ice and snow sublimate directly into water vapor, and additional water is released from flora and fauna via transpiration and sweating respectively. The density of water vapor is lower than the oxygen and nitrogen in the atmosphere and causes the vapor to rise. As it rises into low pressure areas of the atmosphere, it expands and releases energy and cools down. Subsequently, the water vapor condenses into miniscule water droplets on dust and snowflakes in the atmosphere. These droplets then coalesce together and at some point, become heavy enough to come down as precipitation due to gravity. The precipitation then replenishes the various bodies of water on both Earth's surface and underground. The water that falls on land eventually makes its way into the oceans by surface or underground runoff, thus completing the water cycle.

This simple model's elegant explanation hides a very complicated reality that science is still exploring. The precise mechanics of how a cloud forms and grows is not completely understood, but scientists have developed theories explaining the structure of clouds by studying the microphysics of individual droplets. While we know that condensation droplets coalesce together to form rain droplets, we do not know how this coalescence happens. Likewise, the numerous intricate pathways and processes by which rain water traverses the overland and subterranean lithosphere leaves much to be discovered and learned. Even with all our modern technology we have limited understanding of precipitation patterns.

Aruṇapraṣna

Rain has a significant place in the *Vedic* world. Apart from *Sūktas* in the *Veda* that are dedicated to *Parjanyā* (पर्जन्यः) *Devatā*, there are numerous *Sūktas* to other *Devatā*s including *Indra, Agni, Varuṇa, Mitra* (मित्रः), the *Marutas* (मरुताः), and the *Aṣvins* (अश्विनौ) that talk about rainfall. The *Nirūḍapaṣubanḍa* (निरूढपशुबन्ध) *Ṣrota* ritual and one of the *Cāturamāsya* (चातुरमास्य) *Ṣrota* rituals, specifically the *Varuṇapragāsa* (वरुणप्रघास), are performed in the rainy season. Rainfall is one of the beneficial outcomes of most *Yajṇas*. The Ramayana, Mahabharata and Puranas have numerous references to *Yajṇas* for rain. Several *non-Yajṇic* practices such as *Vratas* (व्रत) are prescribed specifically during rainy periods. *Āyurveda* has specific diet and lifestyle guidance for the rainy season. *Cāṇakya's* (चाणक्य) *Artaṣāstra* (अर्थशास्त्र) talks about this subject as well. Classical *Saṁsakṛtam* (संसकृतम्) literature is replete with beautiful descriptions of the same, especially *Kālidāsa's* (कालिदास) *Ṛtusaṁhāra* (ऋतुसंहार).

Aruṇapraṣna, being a part of *Veda*, talks about rainfall within the context of *Yajṇa*. While it does not attempt to explain the science behind rainfall itself, scientific analysis of portions of the *Aruṇapraṣna* shed light on the understanding of *Vedic* ancestors on this subject. In this chapter, I will attempt to scientifically interpret interrelated concepts about rainfall through 3 postulates based on the relevant references in *Aruṇapraṣna*, which I outline below.

Aruṇapraṣna Varṣā Mantras

Prapāṭaka 1 *Anuvākā* 9 *Mantra* 14

अथ निगद॑व्याख्या॒ताः ।

ताननुक्र॑मिष्या॒मः ।

After having offered *Stuti* (स्तुति) to *Sarasvatī* (सरस्वती), we now describe systematically the greatness of *Vāyu*.

Prapāṭaka 1 *Anuvākā* 9 *Mantra* 15

वराहवंस्स्वतपसः ।

विद्युन्मंहसो धूपंयः ।

श्वाप्यो गृहमेधाँश्चेत्येते ।

ये चेमेऽशिंमिविद्विशः ।

From *Varāhava* (वराहव) to *Gṛhameḍa* (गृहमेध) are the 6 *Vāyu* Gaṇas (वायु गणाः) that cause the rains. Each of the *Vāyu Gaṇas* is a set of *Vāyus*. *Varāhava* - *Vāyus* that brings about best type of rainfall, *Svatapasa* (स्वतपसः)- *Vāyus* that are self-energized, *Viḍyunmahasa* (विध्युन्महसः)- *Vāyus* who are as brilliant as lightning, *Ḍūpaya* (धूपयः)- *Vāyus* which bring fragrance in all objects, *Ṣvāpaya* (श्वापयः)- *Vāyus* that spread quickly, *Gṛhameḍa* - *Vāyus* that reside in the homes of *Nityāgnihotris* and stimulate the intellect of the residents of the house. These 6 *Vāyus* are together. There is a seventh *Vāyu*, *Aṣmividviṣa* (अश्मिविद्विशः) who is beneficial to crops. This is *Nāmaḍeya* (नामधेय). They are all described in detail later.

Prapāṭaka 1 *Anuvākā* 9 *Mantra* 16

पर्जन्यास्सप्त पृथिवीमभिवंर्षन्ति ।

वृष्टिंभिरिति ।

एतयैव विभक्तिविंपरीताः ।

सप्तभिर्वातैंरुद्रीरिताः ।

अमूँल्लोकानभिवंर्षन्ति ।

तेषांमेषा भवंति ।

The seven kind of *Vāyus* described earlier create 7 kinds of clouds which cause rain on Earth. The amount of rainfall in any place is decided by these *Vāyu Gaṇas* as per requirements/desired of that place. These same *Vāyu Gaṇas* cause rain in all the above *Lokās* (लोकाः) as well.

Prapāṭaka 1 *Anuvākā* 9 *Mantra* 17

स॒मा॒नमे॒तदु॒द॑कम् ।

उ॒च्चैत्य॑व॒चाह॑भिः ।

भूमिं॑ प॒र्जन्या॒ जिन्व॑न्ति ।

दिवं॑ जिन्वन्त्यग्न॑य इ॒ति ।

Even though the *Ḋravyasvaḇāva* (द्रव्यस्वभाव) or composition of the water that comes down to earth as rain is same as that of the water that resides in the upper worlds, some days it comes down as rain and some days it goes up. The clouds that cause rain on Earth are called *Parjanyās* and the clouds that cause rain in the upper worlds are called *Agneyās* (अग्नेयाः).

Prapāṭaka 1 *Anuvākā* 10 *Mantra* 5

त्युग्रो॑ह भु॒ज्युम॑श्विनोदमे॒घे ।

र॒यिन्न कश्चि॑न्ममृ॒वां २ आवा॑हाः ।

तमू॑हथुर्नौ॒भिरा॑त्म॒न्वती॑भिः ।

अ॒न्त॒रि॒क्ष॒प्रुड्भि॒रपोदकाभिः ।

Oh *Asvins*, just as a foolish person even when dying accumulates wealth without giving it to others, collect sustenance waters from the clouds and bring them to us. These waters should be such that we should be able to travel into the *Antarikṣa* (अन्तरिक्ष) in a well-formed leak free boat.

Prapāṭaka 1 *Anuvākā* 10 *Mantra* 6

ति॒स्रः क्षप॒स्त्रिरहा॑ऽति॒व्रज॑द्भिः ।

नास॑त्या भु॒ज्युमू॑हथुः पत॒ङ्गैः ।

स॒मु॒द्रस्य॒ धन्व॑न्ना॒र्द्रस्य॑ पा॒रे ।

त्रि॒भीरथै॑श्श॒तप॑द्भि॒ः षड॑श्वैः ।

Oh *Asvins*, who are lacking any falsehood, riding on a chariot driven by 6 horses, traveling for 3 days and 3 nights bring us bountiful sustenance waters from the oceans to the dry places.

Prapāṭaka 1 *Anuvākā* 10 *Mantra* 7

स॒वि॒तारं॒ वित॑न्वन्तम् ।

अनु॑बध्नाति शाम्ब॒रः ।

आपपूरुषम्ब॑रश्वै॒व ।

स॒विता॑ऽरेप॒सो॑ भवत् ।

Clouds that hold *Ṣāmbara* (शाम्ब॒रः) waters bind with sun to produce sustenance rains, which are free of any harmful properties.

Varṣā Postulates

Varṣā Postulate 4.1 - Water that falls as rain has quantum nutritious properties that are yet to be understood.

Varṣā Postulate 4.2 - *Vāyu Gaṇas* described in *Aruṇapraṣna* 1.9.15 - 1.9.16 can be understood as quantum fields that helps create condensation.

Varṣā Postulate 4.3 - As per *Aruṇapraṣna* 1.9.17, there is an upward showering precipitation, which is a counterpart to the normally observed downward falling rain.

Isolated Towering Vertical Thunderhead in the Mojave Desert.

Varṣā Postulate 4.1

Water that falls as rain has quantum nutritious properties that are yet to be understood.

The commonly understood concept of rain is that it is water falling from the sky. However, the source, contents and effects of rainfall are variegated. Its characteristics depend on various complexly interrelated factors. My first postulate is that, *Aruṇapraṣna* provides a window into this line of thinking.

The source, content and effect of rainfall is different in various situations. We can categorize the source of rainfall as natural and manmade. The natural rainfall results from only the water-cycle driven precipitation. In *Vedic* parlance, this is considered as *Yajña* inherent in the Nature. Sometimes, human actions alter the natural water-cycle, which can lead to a negative or positive consequence. For example, excessive pollution has resulted in acid rain in several parts of the world. On the other hand, humans can positively influence rainfall as well. The *Vedic Yajña* complex was one such vehicle through which quantum entanglement was created to influence the rain process. Humans formost part do not live off nature but alter it and grow crops to survive. This means nutritious rains are essential, so humans have to mimic nature's *Yajña*. Scientists and atmospheric engineers have also been able to seed cloud and create rain. This manmade rain can be understood as a necessary part of securing water for advanced domestication of plants.

Now coming to content, as we discussed in the Earth chapter, water from all sources is not the same. My hypothesis based on *Aruṇapraṣna* is that rain water has special quantum properties that are essential for healthy sustenance of earth's flora and fauna. *Aruṇapraṣna* implies that rain water is unique in that it is purified and seeded with certain nutritional properties. My hypothesis is that these properties result from certain quantum processes occurring in the atmosphere. This is inferred from discussion in *Aruṇapraṣna* 1.10.7 about *Ṣambara Udaka* (शम्बर उदक) within clouds that binds with sunlight to create rain. We must understand that sunlight's influence in this context of *Ṣambara Udaka* is not meant as the one that leads to the formation of the clouds via evaporation-transpiration processes. Instead, it alludes to yet to be understood quantum interactions between clouds and sunlight. In *Aruṇapraṣna* 1.10.5, the Gods of health, *Aṣvins* are requested to bring the sustenance waters for earth's inhabitants. My hypothesis is that *Aṣvins* here represent the nutrition aspects of the rain water. I contend that we still do not fully understand the quantum attributes of the water that comes down as rain. Scientists today are studying the quantum processes within photosynthesis and possibly these quantum attributes of rain water play a role in those processes. So, the food produced from rain watered plants is healthier than the one that is from the plants watered with other sources. In *Aruṇapraṣna* 1.10.6, a timeline of three days and three nights is indicated for these nutrition inducing processes in the atmosphere to work and produce these sustenance rains. This is to be understood with the metaphorical reference to the travel time of the *Aṣvins* in that *Mantra.*

Vāyu Ganas

Condensation Quantum Fields.

वराहव	**Varāhava**
स्वतपस	**Svat́apasa**
विध्युन्महस	**Vid̲́yunmahasa**
धूपय	**D̲́ūpaya**
श्वापय	**S̲vāpaya**
गृहमेध	**Gr̲hamed̲́a**
अश्मिविद्विश	**As̲mivid́vis̲a**

Vāyugaṇas, Condensation Quantum Fields.

Varṣā Postulate 4.2

Vāyu Ganas are condensation quantum fields.

The second concept related to rainfall is exploration of cloud physics. In *Aruṇapraṣna* 1.9.15 and 1.9.16, there is description of seven types of *Vāyu Gaṇas*. A *Gaṇa* means a group and can also be interpreted as an energy field or some other type of field. As per *Aruṇapraṣna*, these seven *Vāyu Gaṇas* are instrumental in producing the rains. *Aruṇapraṣna* defines these *Vāyu Gaṇas* in terms of their properties as they pertain to rainfall they produce.

Varāhavas, the first *Vāyu Gaṇas*, is a group of *Vāyus* defined as one which bring the best type of rainfall. From where they bring is not discussed in *Aruṇapraṣna*, but we can interpret the source as certain quantum fields from which these *Vāyus* manifest these nutritious rain causing quantum particles in the atmosphere. *Svatapasa* are the *Vāyus*, which are self-energized. As per *Sāyaṇācārya*, these fields are not dependent on sun, fire, or any other source of energy. It could be that certain spontaneous quantum energy fields are formed during cloud condensation process. Alternatively, these maybe condensation causing quantum energy fields that always exist in the atmosphere. *Vidyunmahasa*, are the *Vāyus* that are as brilliant as lightning. *Ḍūpaya* are the *Vāyus* that bring fragrance to all objects. *Ṣvāpaya Vāyus* are self-propagating and spread quickly. *Gṛhameḍa* are the *Vāyus* that reside in the homes of *Nityāgnihotris* and stimulate the intellect of the residents of the house. The seventh *Vāyu Gaṇa* are called *Aṣmividviṣa*, which are described as conducive to agricultural activities. I hypothesize that these seven *Vāyu Gaṇas* are seven kinds of interacting quantum fields. When these fields collapse into their particle states, they cause cloud droplet coalescence. The behavior of these particles is driven by quantum physics, which in conjunction with the water cycle causes condensation. As per *Aruṇapraṣna* 1.19.16, there are seven kinds of clouds corresponding to the seven kinds of *Vāyu Gaṇas*. So, each type of *Vāyu Gaṇa* field collapses to result in a specific type of cloud condensation, which then produces a rainfall that results in a specific benefit. This means all rains are not the same and each rainfall produces a unique impact, especially on the biosphere. It is most likely that a given rainfall is a result of a combination of multiple interacting field collapses and the consequent formation of a complex cloud system. These *Gaṇas* may be responsible for not only cloud droplet coalescence but many other aspects of cloud physics that needed to be studied further. More research is required in the direction of applying quantum theory to cloud micro physics to shed light on these *Vāyu Gaṇas.*

Aerial view of the Amazon Rainforest, near Manaus, the capital of the Brazilian state of Amazonas.

Varṣā Postulate 4.3

There is upward showering rainfall

Rainfall is normally believed to be something that comes down from the skies. However, *Aruṇapraṣna* 1.9.17 talks about a yet to be discovered concept of rainfall showering upwards in the atmospheric layers. *Aruṇapraṣna* states that even though the composition, *dravyasvaḅāva* (द्रव्यस्वभाव) of the water that falls as rain is the same as the water that resides in the upper worlds, some days it comes down as rain and some days it showers up. One might think that this is a description of evaporation but it is not because *Aruṇapraṣna* says that it is the clouds that shower downwards on some days and on some days they shower upwards. It also has specific names for the associated clouds. Those that cause rain on Earth are called *Parjanyās* and the clouds that cause rain in the upper worlds are called *Agneyās*.

The concept of raining upwards might seem impossible but we have to look to other such hydro phenomena that might seem likewise. For example, the biggest river on our planet does not flow on the earth's surface but instead it flows in the atmosphere stretching from the Pacific to Atlantic over the Brazilian rainforests. This was discovered only recently by the scientists. Similarly, the concept of upwards showering rain is something science has yet to explore and a deeper study of the greater *Vedic* literature will assist in that effort.

Conclusion

Water is life here on Earth and rainfall is the principal phenomenon through which life-giving water is delivered to plants, animals and humans. Rainfall is not just water falling from the sky as it is endowed through quantum interactions in the atmosphere with certain yet to be understood life sustaining nutrition. Condensation is then driven by interactions amongst 7 types of quantum fields and/or particles, referred to as the *Vāyu* Ganas in *Aruṇapraṣna*.

The potential areas for further research based on this topic include:

- Empirical study of quantum characteristics of rainwater that benefit plant growth beyond the well-established classical understanding of rainwater nitrogen fixation.
- Developing a scientific understanding of *Vāyu Gaṇas* in *Aruṇapraṣna* within the framework of quantum physics.
- Comprehensive study of clouds in the *Vedic* and *Purāṇic* corpus.

Chapter 5
Gurutvākarṣaṇa

Science

Behind the commonly observed natural phenomenon of all objects being pulled down to the earth, lies the mystery of gravity. It is what keeps the atmosphere and oceans from dissipating into space and plays a key role in almost all cycles of nature that sustain life on our planet. Gravity is at play in the earth's revolution around the sun, which along with the earth's axial tilt produces the seasons. It makes the moon revolve around the earth, causing tidal waves and other natural processes dependent on the lunar cycle. Hundreds of years have passed while we try to understand these gravity dependent processes, and there are still numerous more we have yet to discover and learn about.

Scientists today explain the structure of matter using the Standard Model. According to this model, everything in the universe is made up of a few basic building blocks known as fundamental particles, which are governed by three of the four fundamental forces, which include the strong nuclear force, the weak nuclear force, and the electromagnetic force. This model has yet to explain the fourth: gravitational force. There is a hypothesis that posits the existence of a yet undiscovered particle called the graviton, which is supposed to be mediating the force of gravity. Newton's explanation of gravity as a force was good until it failed to explain the perturbations in Mercury's orbit. Then, Einstein's revolutionary idea that gravity was not a force but a consequence of the uneven distribution of mass and energy in the universe came along and replaced the Newtonian explanation. According to Einstein, the uneven distribution of mass and energy in the universe results in a spacetime fabric that defines the path along which all objects in the universe move. To an observer, this gives the perception of an attractive force and a consequent acceleration, but there is no such force that exists. So, the earth, with a smaller mass, is simply following the path around the Sun, which has a much greater mass as defined by the spacetime fabric. This is a result of the distribution of all the mass and energy in the universe including that of the earth and the sun, and not because the sun is pulling at the earth with gravitational force. This spacetime fabric is constantly changing due to the continuous motion of matter and energy in the universe. Sometimes,

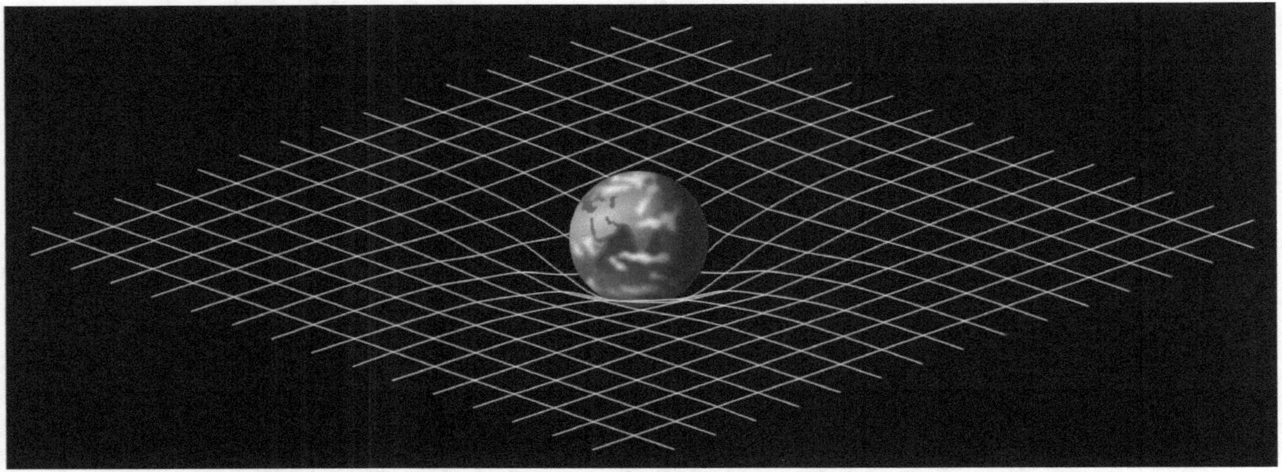

Lattice analogy of the deformation of spacetime caused by a planetary mass.

changes in the spacetime fabric propagate outwards at the speed of light in a wave-like motion. These propagating phenomena are known as gravitational waves. Since 2015, Laser Interferometer Gravitational-Wave Observatories, LIGOs, have been able to measure gravitational waves caused by both the merging of a set of black holes and a set of neutron stars.

To elaborate on what the spacetime fabric is, we must first take a deeper look at time beyond our normal experience in our daily lives. Time can be understood only as a count of repetitions of periodic events. Such events include sunrise-sunset, moon phases, blinking of eyes, and a heartbeat. The electronic transition frequency of the cesium atom is the basis for the international unit of time specifically for the second. Outside of just counting these cyclical repetitions, there is no human understanding of time. Coming back to spacetime, Einstein postulated that time does not pass uniformly for all observers in the universe and it cannot be understood as a mutually exclusive variable from the three-dimensional space. This is because the observed rate at which time passes for an object depends on the object's velocity relative to the observer. So, events that occur at the same time for one observer could occur at different times for another.

To illustrate using a practical example, the clocks on the GPS satellites orbiting in space around the earth run slower by about 0.007 seconds every 6 months compared to the same clocks down on the surface. So, for an observer on earth, a particular time coincides with a specific position of the satellite, which happens simultaneously, but for the other observer those two events happen at different moments. To clarify further, say the observer on earth sees the satellite being over Cape Town, South Africa with the clock reading 10:54:00 AM. This happens simultaneously for the observer on earth, but for the observer on the satellite, the clock there shows approximately 10:53:59.58 AM when it is over the same location. Scientists call this time dilation. This phenomenon has a similar counterpart that can be observed called length contraction, which states that an object's length in motion measures shorter than its length when at rest. So, not only will the observer on earth see time passing quicker relative to the observer on the satellite, but she will also experience the size of the satellite contract by a miniscule amount, altering her read of the satellite's position. So, to describe the motion of an object in space, both space and time must be used in a combination mentioned earlier, spacetime.

Despite Einstein's explanations of gravity through his General Theory of Relativity, numerous questions remain unanswered both within and outside framework of his theory, especially around the concept of quantum gravity. The need for quantum explanations of gravity arises from existence of black holes, dark matter and dark energy where quantum effects cannot be ignored. Einstein's theories and other existing quantum field theories have been inadequate in understanding and solving these mysteries. Scientists are exploring a new frontier of physics to understand these theories and mother nature's gravity enigmas.

Aruṇapraṣna

Aruṇapraṣna has interesting descriptions of time, gravity and the energy fields behind gravity. Its metaphorical representations provide elaborate discussions on these topics, especially if they are read together with associated *Sāyaṇācārya Ḇāṣyas*. In this Chapter, I am proposing three postulates on this subject matter. They are outlined after the section where the related *Aruṇapraṣna Mantras* are presented.

Aruṇapraṣna Gurutvākarṣaṇa Mantras

Prapāṭaka 1 *Anuvākā* 8 *Mantra* 1

क्वेदमभ्रन्निविशते ।

क्वायँ संवत्सरो मिथः ।

क्वाहः क्वेयन्देव रात्री ।

क्व मासा ऋतवः श्रिताः ।

We see the water bearing clouds in the sky without any support. Therefore, the question is on whose support are these clouds staying there? Similarly, the year, day & night and month & season, what are they caused by? The year is composite set of many time subsets.

Prapāṭaka 1 *Anuvākā* 8 *Mantra* 2

अर्धमासा मुहूर्ताः ।

निमेषास्त्रुटिभिस्सह ।

क्वेमा आपो निविशन्ते ।

यदीतो यान्ति संप्रति ।

The perceivable time periods of *Pakṣa* (पक्ष), *Muhūrta* (मुहूर्त), and *Nimeṣa* (निमेष)and the non-perceivable smaller time periods such as *Paramāṇu*

(परमाणु), *Aṇu* (अणु), and *Tṛti* (तृटि) are held by what and from where do they arise? The water that evaporates during hot periods, where does it reside?

Prapāṭaka 1 *Anuvākā* 8 *Mantra* 3

काला अप्सु निविशन्ते ।

आपस्सूर्ये समाहिताः ।

अभ्राण्यपः प्रपद्यन्ते ।

विद्युत्सूर्ये समाहिता ।

This is the answer for the questions in the previous two *Mantras*. All time periods reside in the water. The entire creation came from water as is mentioned in 1.23.1. So, water is where time periods reside and that is where they originate from. Since the sun's rays evaporate the water, it is said to reside in the sun. The clouds are created by a mixture of sun evaporated water, dust, and air and so clouds originate from evaporated water. The lightning resides in the sun. Suns rays only convert into lightning.

Prapāṭaka 1 *Anuvākā* 8 *Mantra* 4

अनवर्णे इमे भूमी ।

इयं चासौ च रोदसि ।

Earth mentioned as *Dvivacana* (द्विवचन)word means earth and sky together. Colorlessness is ugliness and earth and sky are devoid of colorlessness and therefore beautiful. Being devoid of any ugliness and lack of color is beauty. Earth and sky are beautiful. Beauty means they are infused with many *Devatās*.

Prapāṭaka 1 *Anuvākā* 8 *Mantra* 5

किँस्विदत्नान्त॑रा भू॒तम् ।

ये॒नेमे वि॑धृ॒ते उभे ।

विष्णुणा विधृते भूमी ।

इति व॑त्सस्य॒ वेद॑ना ।

What is that which holds the earth and sky? *Vatśa* (वत्स) *Ṛṣi's* statement is that *Viṣṇu* is the one holding earth and sky.

Prapāṭaka 1 *Anuvākā* 8 *Mantra* 6

इरा॑वती धे॒नु॒मती॒ हि भू॒तम् ।

सू॒य॒व॒सिनी॒ मनु॑षे दश॒स्ये᳚ ।

व्य॑ष्टभ्ना॒द्रोद॑सी॒ विष्ण॑वे॒ते ।

दा॒धर्थ॑ पृथि॒वीम॒भितो॑ म॒यूखैः ।

The earth has the capacity to provide bountiful rice, cows, and other beautiful foods to humans. *Viṣṇu* holds the earth and sky. *Viṣṇu* holds them from above with rays.

Prapāṭaka 1 *Anuvākā* 8 *Mantra* 7

किन्तद्विष्णोर्ब॑लमा॒हुः ।

का॒ दीप्तिः॑ किं प॒राय॑णम् ।

एको॑ य॒ध्दार॑यद्दे॒वः ।

रे॒जती॑ रोद॒सी उ॑भे ।

The *Viṣṇu* who alone is holding the earth and sky, where does he get his strength from? The rays of light that *Viṣṇu* uses to hold where does that come from? And who are *Viṣṇu's* helpers?

Prapāṭaka 1 *Anuvākā* 8 *Mantra* 8

वाताद्विष्णोर्बलमाहुः ।

अक्षराद्दीप्तिरुच्यते ।

त्रिपदाध्दारयद्देवः ।

यद्विष्णोरेकमुत्तमम् ।

Viṣṇu produces a thread like substance from *Vāyu* and he uses this thread to hold this whole creation in place. Therefore, *Viṣṇu's* strength and capacity to hold the universe in place comes from *Vāyu*. This is the answer to the question, "where does *Viṣṇu* get his strength from" in the *Mantra* 1.8.7. The rays of light *Viṣṇu* uses to hold the universe is something that never dims, cannot be destroyed, self-created, beyond the material world and it is what lights up the entire universe. This is the answer to the question about the source of the light rays, which *Viṣṇu* uses to hold the universe in *Mantra* 1.8.7.

Prapāṭaka 1 *Anuvākā* 8 *Mantra* 9

अग्नयो वायवश्चैव ।

एतदस्य परायणम् ।

From *Savitara* (सवितर) to *Aruṇaketukāgni* and their associated *Vāyu* friends are *Viṣṇu*'s helpers. This is the answer to the question, "Who are *Viṣṇu*'s friends?" in the *Mantra* 1.8.7.

Gurutvākarṣaṇa Postulates

Postulate 5.1

Viṣṇu in *Aruṇapraṣna* 1.8.1 to 1.8.9 is metaphorical description of gravity.

Postulate 5.2

The *Mayukas* (मयूखाः), *Vāyus* and *Agnis* in *Aruṇapraṣna* 1.8.7 to 1.8.9 are a metaphorical description of a gravitational field, space-time fabric and/or a yet to be discovered quantum gravitational field.

Postulate 5.3

The discussion of time in *Aruṇapraṣna* 1.8.1 to 1.8.3 taken together with space-time fabric like discussion of gravity in *Aruṇapraṣna* 1.8.4 to 1.8.9 is a metaphorical presentation of the time component of the space-time fabric.

Viṣṇu in *Varāha Avatāra*.

Raja Ravi Verma (1848 - 1906), Varaha Avatar (Oil on Canvas), Public Domain.

Gurutvākarṣaṇa Postulate 5.1

Viṣṇu is Gravity

To be accurate, *Aruṇapraṣna* does not broach the subject of gravity directly. It starts this discussion by admiring the beauty of the earth and the heavens, which constitutes of the rest of the universe surrounding earth. It then wonders what holds the heavens and the earth in place and goes on to answer the query saying that it is *Viṣṇu* holding them in place. It is interesting to note that there is no force of attraction discussed in *Aruṇapraṣna*. It presumes there is something keeping the balance, recognizes it, and simply names it *Viṣṇu*. The way *Viṣṇu* holds the universe together is discussed later in that *Anuvākā*. That firmly establishes that *Aruṇapraṣna* has an understanding of gravity, metaphorically ascribed as *Devatā Viṣṇu*. In the greater *Vedic* and *Purāṇic* literature, *Viṣṇu* is described as the sustainer of the creation.

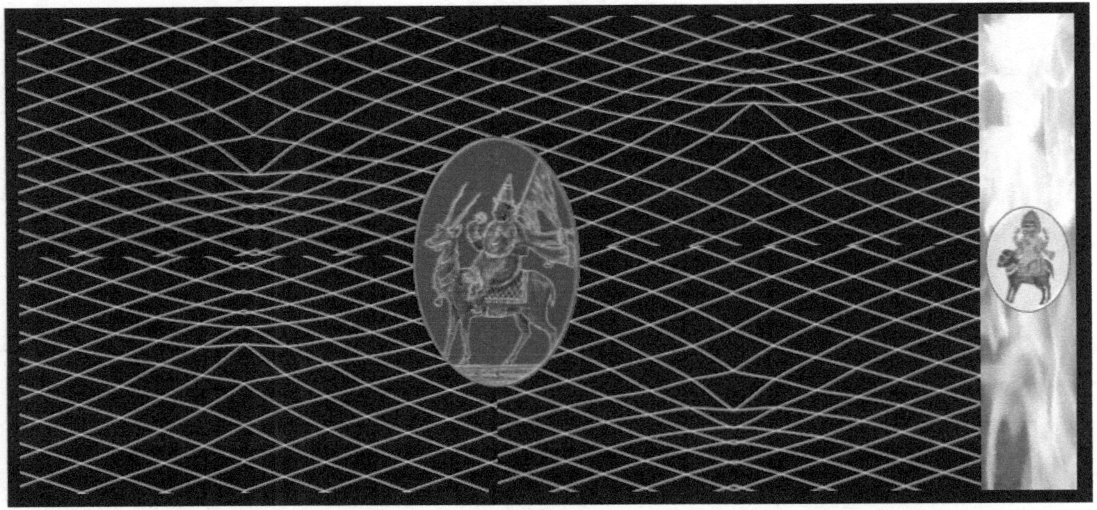

Artist's imagination of Gravity Fields, Space-Time Fabrics or Quantum Gravity Fields as *Vāyu* and *Agni*.

Lattice analogy of the deformation of spacetime caused by a planetary mass., MySid, CC BY-SA 3.0 <https://creativecommons.org/licenses/by-sa/3.0>, via Wikimedia Commons. Agni, god of fire, shown riding a ram, British Museum, Public domain, via Wikimedia Commons. Vayu, The complete Hindoo Pantheon, comprising the principal deities worshipped by the Natives of British India throughout Hindoostan, E. A. Rodrigues, Public domain, via Wikimedia Commons.

Gurutvākarṣaṇa Postulate 5.2

Mayūkas, *Agnis* and *Vāyus* together describe Gravity Fields, Space-Time Fabrics or Quantum Gravity Fields

Aruṇapraṣna states that *Viṣṇu* gets his power from *Agni* and *Vāyu*. *Sāyaṇācārya*, in his *Aruṇapraṣna Bāṣya*, elaborates on the phrase "*Viṣṇu*'s strength comes from *Vāta* (वातः)" in *Aruṇapraṣna* 1.8.8 as *Viṣṇu* producing a thread-like substance from *Vāyu*. This may be a metaphorical description of a wave used by *Viṣṇu* to hold the whole of creation in place. It then goes on to say that various *Agnis* and *Vāyu*s support *Viṣṇu* in his work. *Sāyaṇācārya* continues that these *Agnis* range from *Savitara* to *Aruṇaketukāgni*. *Aruṇapraṣna* discusses various types of *Agnis* and *Vāyu*s throughout its text, even in sections separate from this specific topic. Furthermore, *Aruṇapraṣna* says *Viṣṇu* holds these bodies in place using *Mayūkas*, which *Sāyaṇācārya* describes as an infinite, indestructible and primordial energy source in the universe. *Agnis* can be interpreted as the energy behind these primordial energy rays and *Vāyu* can signify the wave-like characteristics of these rays. Also, the plural *Agnis* and *Vāyu*s posit to there being various fields, not just one. So, as per *Aruṇapraṣna*, there are various fields, space-time fabrics or as of yet unknown quantum force fields that are working together to produce the gravitational effects we observe. This combined effort in *Aruṇapraṣna*'s parlance is defined as *Viṣṇu*.

It is interesting to note that in *Matsya Purāṇa* (मत्स्य पुराण), we find a description of the relative motions of the earth, sun and the Pole Star that signifies a fabric-like structure as well. It uses the potter's wheel to describe the relative motions of these bodies and talks about their bounce during their motions with earth's bounce greater than that of the sun's, and the sun's bounce greater than that of the Pole Star. The potter's wheel is used to describe the motions and wobble of these bodies. The Sun is the clay and the center at the wheel, and the other celestial objects are spread out from that center. The clay weighs down, and pushes the other bodies up around the rest of the wheel. When rotating, it wobbles like the celestial bodies in the universe. The *Purāṇa* gives a second example, with the clay in the center representing the Pole Star instead of the Sun. To me, this appears as a metaphorical description of a space-time fabric.

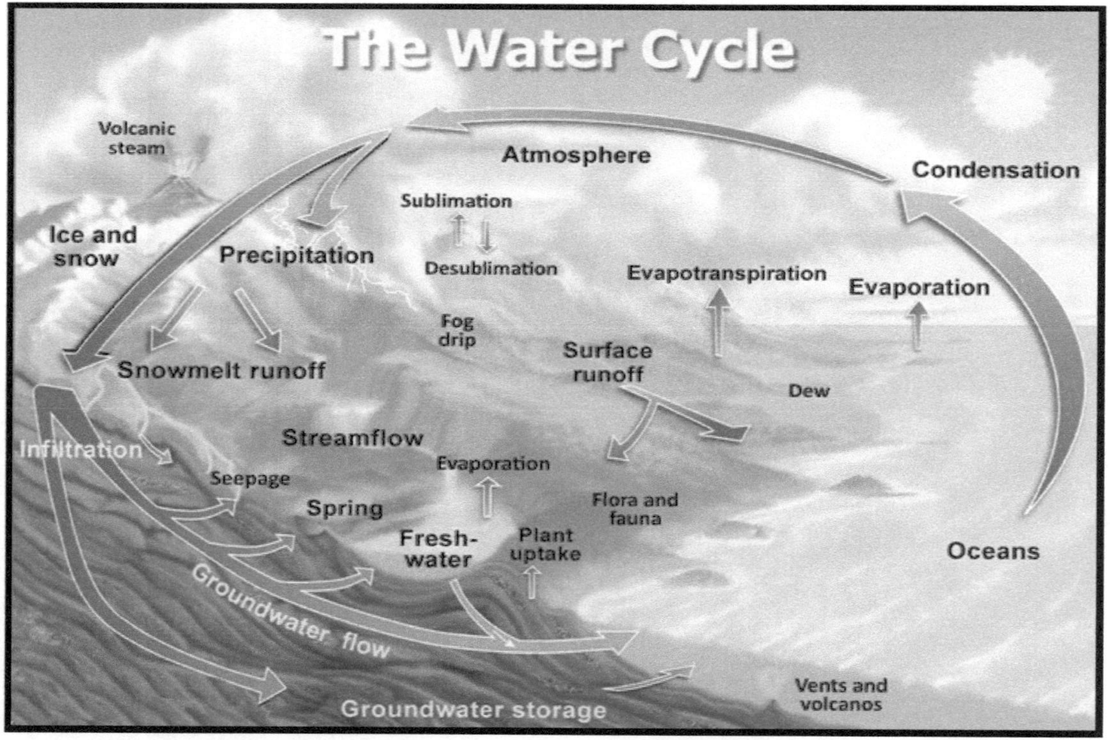

Water-Cycle Diagram.

John Evans and Howard Periman, USGS, Public domain, via Wikimedia Commons.

Gurutvākarṣaṇa Postulate 5.3

Aruṇapraṣna 1.8.1 to 1.8.8 implies Time is part of the Space-Time fabric

Aruṇapraṣna precedes the discussion of gravity with a discussion on time. It focuses on two topics in its discussion: the source of time's origin and the place of its existence. *Aruṇapraṣna* states that water is both the source of time and its place of existence. In the *Būmi* chapter, details on *Aruṇapraṣna*'s description of the origin of earth from water were provided. So, as per *Aruṇapraṣna*, both space and time have their origins in the primordial waters. *Aruṇapraṣna* further states that the water, in which time exists, is integrated within the sun. *Sāyaṇācārya* interprets this integration as a succinct description of the earth's water cycle. It is through the water cycle that time is then manifested within the sun.

The interpretation here is that the Sun does not just refer to the star at the center of the solar system, but also encompasses the rays of light it produces and disseminates. Elsewhere in the same *Anuvākā*, the Sun is called *Paṣyaka* (पश्यक) the one who sees the past, present and future. In the earlier two postulates, we presented *Aruṇapraṣna*'s discussion on how the heavens and the earth are held together by gravity. The heavens include the Sun. The time is manifested in the sun as per *Aruṇapraṣna*. So, *Aruṇapraṣna*'s gravity discussion incorporates time as well. How time is related to gravity is not discussed explicitly in *Aruṇapraṣna* but elsewhere in the greater *Vedic* and *Purāṇic* corpus, there is a more elaborate description of time's role in cosmology and cosmogony of the universe.

Conclusion

Gravity is one of those subjects that science has yet to develop a clear understanding of but plays a fundamental role in the functioning of our world as we know it. In *Aruṇapraṣna*, *Viṣṇu* is shown to hold up the earth, the clouds and the sky. This is a metaphorical description of gravity. *Viṣṇu*'s power is explained to include rays of light, *Agnis* and *Vāyus*. This is a metaphorical description of a gravitational field, space-time fabric and/or yet to be discovered quantum gravitational field. Time originated from the primordial waters from which the rest of creation also emerged. *Aruṇapraṣna*'s discussion of time preceding its description of gravity combined with the description of integration of time within the Sun maybe a description of a space-time like structure.

The potential areas for further research based on this topic include:

- Comprehensive study of references to gravity in the *Vedic* and *Purāṇic* corpus similar to the one described in *Aruṇapraṣna*.
- Comprehensive study of Time in the *Vedic* and *Purāṇic* corpus.

Chapter 6
Sūryaraśmi

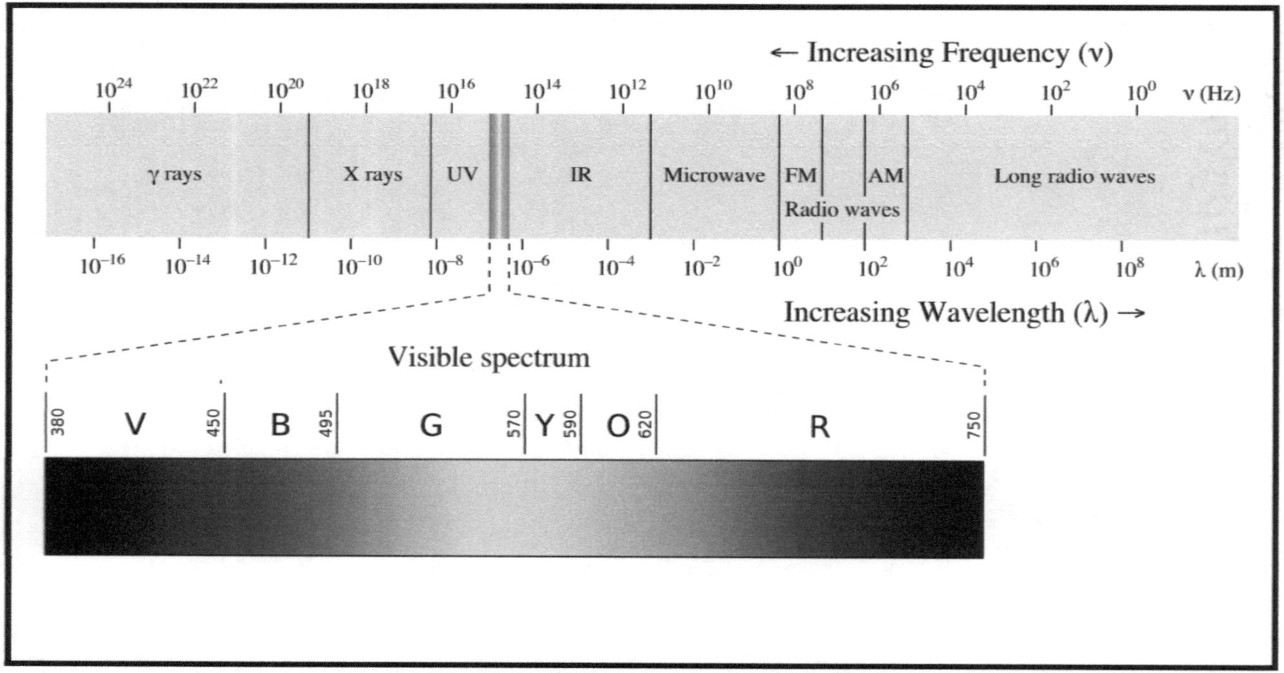

Image showing the distribution of electromagnetic waves with respect to frequency and wavelength, highlighting visible part of the electromagnetic spectrum.

Science

Sunlight, a form of electromagnetic radiation, is the principal source of energy on earth and sustains all life on our planet. The sun provides approximately 1,000 Watts/Square Meter of energy to earth's surface every day. The solar industry harvests a minuscule portion of this energy to produce electricity and is one of the fastest growing industries in the world. The sun's energy is also indirectly used to fuel human activity, as evidenced by the fossil fuel industry that drives the global economy today. Fossil fuels are products of natural processes that decomposed buried organisms that contained energy from ancient photosynthesis processes. But human civilization recognized the importance of the sun to life on earth long before the emergence of these industries. It is no wonder that the sun is considered a deity by almost all indigenous religions of the world. The synodic rotation of the earth and its revolution around the sun are the basis of most solar calendars including the one in use today.

The light that we receive from the sun is a spectrum of electromagnetic radiation. The spectrum consists of waves that range in frequency from just 1 hertz to greater than 300 exahertzs (3×10^{20} hertzs). It is usually divided into 3 groups based on their frequencies.

The first group is the visible part of the spectrum known as visible light. This part of the spectrum is commonly known as sunlight. It stimulates various biological processes including those that underlie human vision and plant photosynthesis. The composition of this visible light is commonly understood in the form of a rainbow, which consists of numerous streams of light that range from red light to violet light.

The first component of the second group is known as infrared light because the frequency of these waves is below that of the visible spectrum's red light. This group also consists of microwaves and radio waves whose frequencies are lower than those of infrared.

Lastly, the third group consists of ultraviolet rays, x-rays and gamma rays in that order of increasing frequencies. The first two groups of the spectrum pass through the earth's atmospheric layers and reach the surface, whereas most of the third group is filtered away by outer layers of the atmosphere.

Aruṇapraṣna

Sūrya is one of the most important *Devatā* in *Veda*. As mentioned earlier in the Introduction chapter, the *Vedic* way of life and the accompanying mandatory rituals follow the solar cycles, lunar cycles and the practitioner's biological life cycle. The sun plays a crucial role in all the ancillary *Vedic* sciences including *Yoga* (योग), *Āyurveda* (आयुर्वेद), *Vāstuṣāstra* (वास्तुशास्त्र), and *Bāratīya* Astrology. The *Ādityahṛdayam* (आदित्यहृदयम्) of *Rāmāyaṇā* (रामायणा), *Sūryāṣṭottarasatanāmāvalī* (सूर्याष्टोत्तरशतनामावली) of *Mahābārata* (महाभारता) and the descriptions of various solar dynasties in the *Purāṇas* are testimonies of the importance of *Sūrya in the greater Vedic tradition.* In this chapter, I am proposing 3 postulates on this subject matter. They are outlined after the section below where the related *Aruṇapraṣna Mantras* are presented.

Aruṇapraṣna Sūryaraṣmi Mantras

Prapāṭaka 1 *Anuvākā* 7 *Mantra* 1

आरोगो भ्राजः पतरः पतङ्गः ।

स्वर्णरो ज्योतिषिमान् विभासः ।

ते अस्मै सर्वे दिवमातपुन्ति ।

ऊर्जं दुहाना अनपस्फुरन्त इति ।

Āroga (आरोग), *Brāja* (भ्राज), *Patara* (पतर), *Pata˙ga* (पतंग), *Svarṇara* (स्वर्णर), *Jyotiṣmān* (ज्योतिष्मान्) and *Vibāsa* (विभास) are the names of seven forms of the *Sūrya Devatā* (सूर्य देवत). They light up the sky and the heavens for the benefit of our world. Through rain they provide milk and other nourishment. They do not emit rays that are harmful for life forms. So, they only shine that which is good for living beings. These seven *Sūryas* illustrate the bricks used for the *Aruṇaketukāgni Yajṇa*. From the *Mantra* we can infer that these *Saptasūryas* (सप्त सूर्याः) are like bright *Dṛva* (धृव) star.

Prapāṯaka 1 *Anuvākā* 7 *Mantra* 2

कश्य॑पोऽष्ट॒मः ।

स महा॒मे॑रुं न॑ जहा॒ति ।

तस्यै॒षा भव॑ति ।

The eighth sun, which is separate from the seven discussed in *Mantra* 1.7.1 is called *Ksyapa* (क्श्यप). He never leaves *Mahameru* (महामेरु) mountain or in other words he always shines on *Mahameru* mountain. More will be discussed about this *Ksyapa Surya* in the coming *Mantra*s.

Prapāṯaka 1 *Anuvākā* 7 *Mantra* 3

यत्ते॒ शिल्पं॑ कश्यप रोच॒नाव॑त् ।

इन्द्रि॒याव॑त्पुष्क॒लं चि॒त्रभा॑नु ।

यस्मि॒न्सूर्या॒ अर्पि॑तास्स॒प्त सा॒कम् ॥

तस्मिन्राजानमधिविश्रिये॑ममि॒ति ।

Ksyapa, your splendorous rays are radiating the world with light, are capable of sensory perception, contain various kinds of rays and with which the *Saptasuryas* (*Aroga* etcetera) are always together. Using these rays bring light to the *Yajamana* (यजमान).

Prapāṯaka 1 *Anuvākā* 7 *Mantra* 4

ते अस्मै सर्वे॑ कश्यपाज्ज्योति॑र्मभ॒न्ते ।

तान्सोमः कश्यपादधि॑निर्ध्द॒मति ।

भ्रस्ताकर्म॑कृदिवै॒वम् ।

To provide light to the world, the *Saptasuryas* (*Aroga* etcetera) obtain the light from *Ksyapa*. Even though they possess the energy to light the world, *Soma Devata* (सोम देवत) purifies and brightens their light utilizing the light of *Ksyapa* just as a goldsmith uses the air in the bellows to create the fire to purify and brighten the gold.

Prapāṭaka 1 *Anuvākā* 7 *Mantra* 5

प्राणो जीवानीन्द्रियंजीवानि ।

सप्त शीर्षण्याः प्राणाः ।

सूर्या इंत्याचार्याः ।

According to various *Ācāryas* (आचार्याः), the *Saptasūryas* are: Breath that moves through the mouth and nose in its seven types or *Mahata* (महत), *Ahaṅkāra* (अहङ्कार) and the five *Tanmātras* (तन्मात्राः) or *Manasa* (मनस), *Buddi* (बुद्धि) and the five *Indriyas* (इन्द्रियाः) or The orifices in the head, namely the two eyes, two ears, two nostrils and the mouth.

Prapāṭaka 1 *Anuvākā* 7 *Mantra* 6

अपश्यमहमेतन्सप्त सूर्यानिति ।

पञ्चकर्णो वात्स्यायनः ।

सप्तकर्णश्च प्लाक्षिः ।

आनुश्राविक एव नौ कश्यप इति ।

उभौ वेदयिते ।

न हि शेकुमिव महामेरुं गन्तुम् ।

Pañcakarṇa (पञ्चकर्ण), son of *Vatsa Ṛṣi* and *Saptakarṇa* (सप्तकर्ण), son of *Plākṣi* (प्लाक्षि) *Ṛṣi* both saw the *Saptasūryas* but they never saw *Kṣyapa*, about whom they only learnt from their respective teachers. They said they are not capable of going to the Mount *Mahāmeru* where *Kṣyapa* resides.

Prapāṭaka 1 *Anuvākā* 7 *Mantra* 7

अपश्यमहमेतत्सूर्यमण्डलं परिवर्तमानम् ।

गार्ग्यः प्राणत्रातः ।

गच्छन्त महामेरुम् ।

एकं चाजहतम् ।

Garga (गर्ग) *Ṛṣi's* son *Prāṇatrāta* (प्राणत्रात) *Ṛṣi* said to *Ṛṣis Saptakarṇa* and *Pañcakarṇa* that he has seen *Kṣyapa*. So go to *Mahāmeru* and you can see *Kṣyapa*.

Prapāṭaka 1 *Anuvākā* 7 *Mantra* 8

भ्राजपतरपतंङ्गा निहने ॥

तिष्ठन्नांतपन्ति ।

तस्मांदिह तप्विंतपाः ।

The center of the *Saptasūryas*, *Āroga* is seen by everyone. The *Sūryas*, *Brāja*, *Patara* and *Pata˙ga* are located below the path to *Mahāmeru* and the light from these *Sūryas* shines on earth. The other three *Sūryas*, namely *Svarṇara*, *Jyotiṣmān* and *Vibāsa* are located above the path to *Mahāmeru* and the light from these *Sūryas* does not shine on earth. Between these two *Sūrya* sections one can find the path to travel to *Mahāmeru* to see *Kṣyapa*.

Prapāṭaka 1 *Anuvākā* 7 *Mantra* 9

तेषांमेषा भवंति ।

This *Mantra* is about those *Saptasūryas*.

Prapāṭaka 1 *Anuvākā* 7 *Mantra* 10

सप्त सूर्या दिवमनुप्रविंष्टाः ।

तानन्वेतिं पथिभिंर्दक्षिणावान् ।

ते अस्मै सर्वे घृतमांतपन्ति ।

ऊर्जं दुहाना अनपस्फुरन्त इति ।

The *Saptasūryas* are systematically spread-out in the heavens. Using the water bricks and correct *Dākṣiṇa* (दक्षिण), the properly done *Aruṇaketukāgni* ritual makes the *Aruṇaketukāgni* to travel between the aforementioned middle path in *Aruṇapraṣna* 1.7.8 to *Mahāmeru* and bring back milk and other things (the fruits of the *Yajña*) for the *Yajamāna.*

Prapāṭaka 1 *Anuvākā* 7 *Mantra* 11

सप्तर्त्विजः सूर्यो इत्याचार्याः इति ।

According to some *Ācāryas*, *Ṣrōta Yajña* priests are the *Saptasūryas*.

Prapāṭaka 1 *Anuvākā* 7 *Mantra* 12

तेषामेषा भवति ।

This *Mantra* is about those *Saptasūryas*.

Prapāṭaka 1 *Anuvākā* 7 *Mantra* 13

सप्त दिशो नानासूर्याः ।

सप्त होतार ऋत्विजः ।

देवा आदित्या ये सप्त ।

तेभिः सोमाभीरक्षण इति ।

Among the eight directions, east is well known as *Sūrya's* direction. The remaining seven directions also have their respective *Sūryas*. Those *Saptasūryas* exist in the form of seven kind of *Ṣrōta* Priests. They are *Hotā* (होता), *Prṣāstā* (प्रशास्ता), *Brahmaṇācansi* (ब्रह्मणाच्छन्सि), *Potā* (पोता), *Neṣṭa* (नेष्ट), *Accāvāka* (अच्छावाक), *Āgnīdra* (आग्नीध्र). These seven priests becoming *Devatās* are called the seven *Ādityas* (आदित्याः). *Soma Devatā* along with those seven *Ādityas* protect us.

Prapāṭaka 1 *Anuvākā* 7 *Mantra* 14

तदंप्याम्नायः ।

दिग्भ्राज ऋतूंन् करोति ।

The *Sūryas* in the other seven directions also produce the seasons in those seven directions.

Prapāṭaka 1 *Anuvākā* 7 *Mantra* 15

एतयैवावृता सहस्त्रसूर्यताया इति वैंशंपायनः ।

According to *Veśa'pāyana* (वैशंपायन) *Ṛṣi* there are infinite *Sūryas* in various directions and therefore it should not be surprising if we say that there are *Saptasūryas*.

Prapāṭaka 1 *Anuvākā* 7 *Mantra* 16

तस्यैषा भवंति ।

This *Mantra* is about those infinite *Sūryas*.

Prapāṭaka 1 *Anuvākā* 7 *Mantra* 17

यद्द्यावं इन्द्र ते शतँशतं भूमिं ।

उत स्युः ।

नत्वां वज्रिन्सहस्त्रँसूर्याः

अनुनजातमष्ट रोदंसी इति ।

Vajra (वज्र) wielding *Indra*, between your infinite heavens and infinite earths, the infinite *Sūryas* that exist do not encompass you. This is some way related to the *Aruṇaketukāgni* Bricks as well.

Prapāṭaka 1 *Anuvākā* 7 *Mantra* 18

नानालिङ्गत्वादृतूनां नानासूर्यत्वम् ।

The variegated characteristics of seasons provide the reason for existence of many *Sūryas*.

Prapāṭaka 1 *Anuvākā* 7 *Mantra* 19

अष्टौ तु व्यवसिता इति ।

The *Ṛṣis* having seen the eight suns with their divine eyes have established that fact, that there are eight *Sūryas*.

Prapāṭaka 1 *Anuvākā* 7 *Mantra* 20

सूर्यमण्डलान्यष्टात ऊर्ध्वम् ।

There are eight *Sūrya Maṇdalas* (मण्डलाः) above. The *Ṛṣis* with divine sight can see these eight *Maṇdalas* whereas we with normal sight can only see one sun.

Prapāṭaka 1 *Anuvākā* 7 *Mantra* 21

तेषामेषा भवति ।

This *Mantra* provides knowledge about these other many *Sūrya Maṇdalas*.

Prapāṭaka 1 *Anuvākā* 7 *Mantra* 22

चित्रं देवानामुदगादनीकम् ।

चक्षुर्मित्रस्य वरुणस्याग्नेः ।

आप्रा द्यावापृथिवी अन्तरिक्षम् ।

सूर्य आत्मा जगतस्तस्थुषश्चेति

He, who is mysterious because of having the form of many *Maṇdalas*, is the destroyer of the enemy in the form of darkness, is like an army of the *Devatās*, rises in the east and in whom *Mitra*, *Varuṇa*, and *Agni Devatās* are located. He provides light to the heavens, the space between heavens and earth and the earth is the *Ātma* (आत्म) of all fixed and moving things.

Sūryaraṣmi Postulates

Postulate 6.1

The *Saptaśūryas* discussed in *Aruṇapraṣna* 1.7.1 through *Aruṇapraṣna* 1.7.22 are metaphorically the seven major components of the electromagnetic radiation spectrum.

Postulate 6.2

Aruṇapraṣna 1.7.4 is a metaphoric presentation of a yet to be discovered process, by which the solar electromagnetic spectrum undergoes a signal enhancement.

Postulate 6.3

The classification of the *Saptaśūryas* into ones that shine on earth and the ones that don't in *Aruṇapraṣna* 1.7.8 is a metaphoric division of the electromagnetic spectrum into radiations that reach the earth's surface and the ones that are absorbed or reflected by the earth's various atmospheric layers.

Sapta Sūryas **EMR Spectrum**			
भ्राज	**Bṟāja**	**Reaches Earth**	**Not Visible**
पतर	**Patara**	**Reaches Earth**	**Not Visible**
पतंग	**Pata˙ga**	**Reaches Earth**	**Not Visible**
अरोग	**Aroga**	**Reaches Earth**	**Visible**
स्वर्णर	**Svarṇara**	**Does not Reach Earth**	**Not Visible**
ज्योतिष्मान्	**Jyotişmān**	**Does not Reach Earth**	**Not Visible**
विभास	**Vibāsa**	**Does not Reach Earth**	**Not Visible**

Saptasūryas, EMR Spectrum.

Sūryaraṣmi Postulate 6.1

Saptasūryas are EMR Spectrum

The entire seventh *Anuvākā* of *Aruṇapraṣna* is a discussion centered around the concept of the *Saptasūryas*. According to *Aruṇapraṣna*, there are *Saptasūryas*: *Āroga*, *Ḅrāja*, *Paṭara*, *Paṭa'gaḥ*, *Svarṇara*, *Jyotiṣmān*, *Viḅāsa*. My hypothesis is that this is a metaphoric representation of the *Aruṇapraṣna's* classification of the electromagnetic spectrum. This classification is different than the scientific one, which classifies based on frequencies into categories namely, the radio waves, microwave radiation, infrared radiation, visible light, ultraviolet radiation, x-ray radiation and gamma radiation. More research into the greater *Vedic* corpus is needed to establish a one-to-one correlation between the components of the electromagnetic spectrum and the individual *Saptasūryas*. What we can say for sure is though that *Aruṇapraṣna recognizes the components of the electromagnetic spectrum and classifies them.*

Aruṇapraṣna discusses a couple of related concepts. First, there is a mention of infinite *Sūryas*. This may indicate that there are various sub frequencies of the seven broad classifications of the electromagnetic spectrum. It also says that even though there are these innumerable *Sūryas*, they do not penetrate the earth's atmospheric protection layers, which we identified earlier to be *Indra* and other *Devatās*. There is a detailed discussion on this concept in the Atmospheric layers chapter. Secondly, it says that the variegated characteristics of the seasons provide empirical evidence to prove that there are multiple *Sūryas*. More research into *Veda* and also Science is needed to understand what *Aruṇapraṣna* means by these assertions.

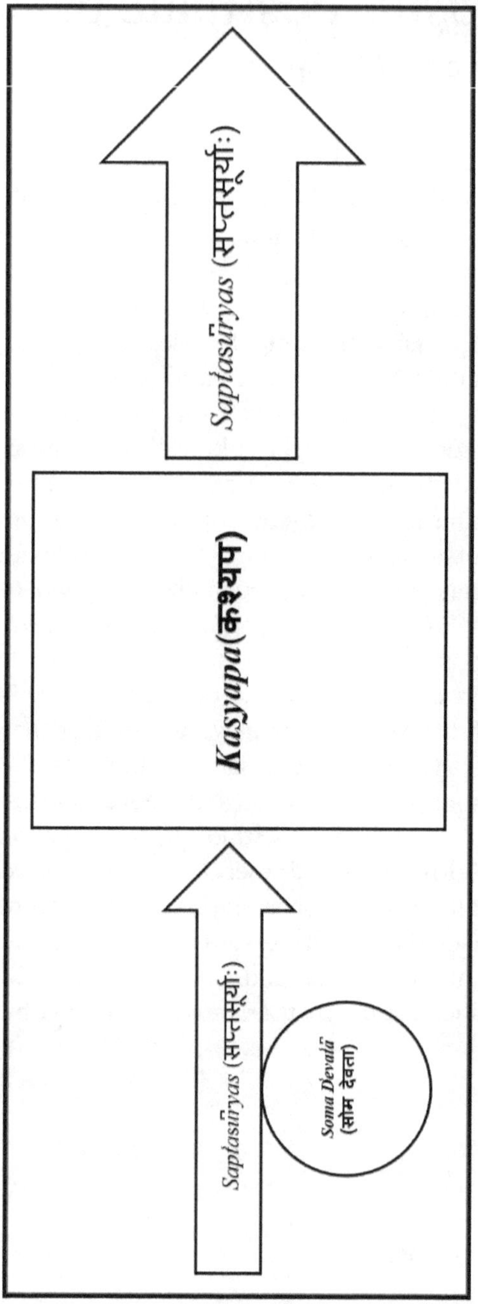

Soma enhances *Saptasūryas*.

Sūryaraşmi Postulate 6.2

Soma is a yet to be discovered field that enhances the *Saptásūryas* (EMRs)

Aruṇapraşna also talks about an eighth radiation called *Kşyapa*, which it says is only perceivable on a mountain referred to as *Mahāmeru*. *Aruṇapraşna* presents an interesting conversation amongst *Ṛşis* that clarifies the difficulty in reaching *Mahāmeru* and observing *Kşyapa*. It talks about a path through the *Saptásūryas* that will lead one to the spot where *Kşyapa* can be observed. But more importantly, *Aruṇapraşna* says that the energy of the *Saptásūryas* is accentuated by the *Kşyapa*. *Kşyapa* does not produce these *Saptásūryas* but merely enhances their capacity so that they can serve the needs of the world. An additional variable referred to as *Soma*, another *Vedic* deity, is introduced into this discussion as the carrier of these *Saptásūryas* to *Kşyapa* for the purpose of energizing and increasing their brilliance, just as a goldsmith polishes gold and enhances its beauty.

So, what are *Soma* and *Kşyapa* metaphorically representing? Is this a discussion to describe the inner workings of an electromagnetic radiation wave propagation? Is *Soma* a magnetic or another yet to be discovered field? Or, is *Kşyapa* the sun and *Soma* a field in the sun that enhances the EMR prior to discharging the EMRs into the solar system and beyond? These questions motivate us towards more research.

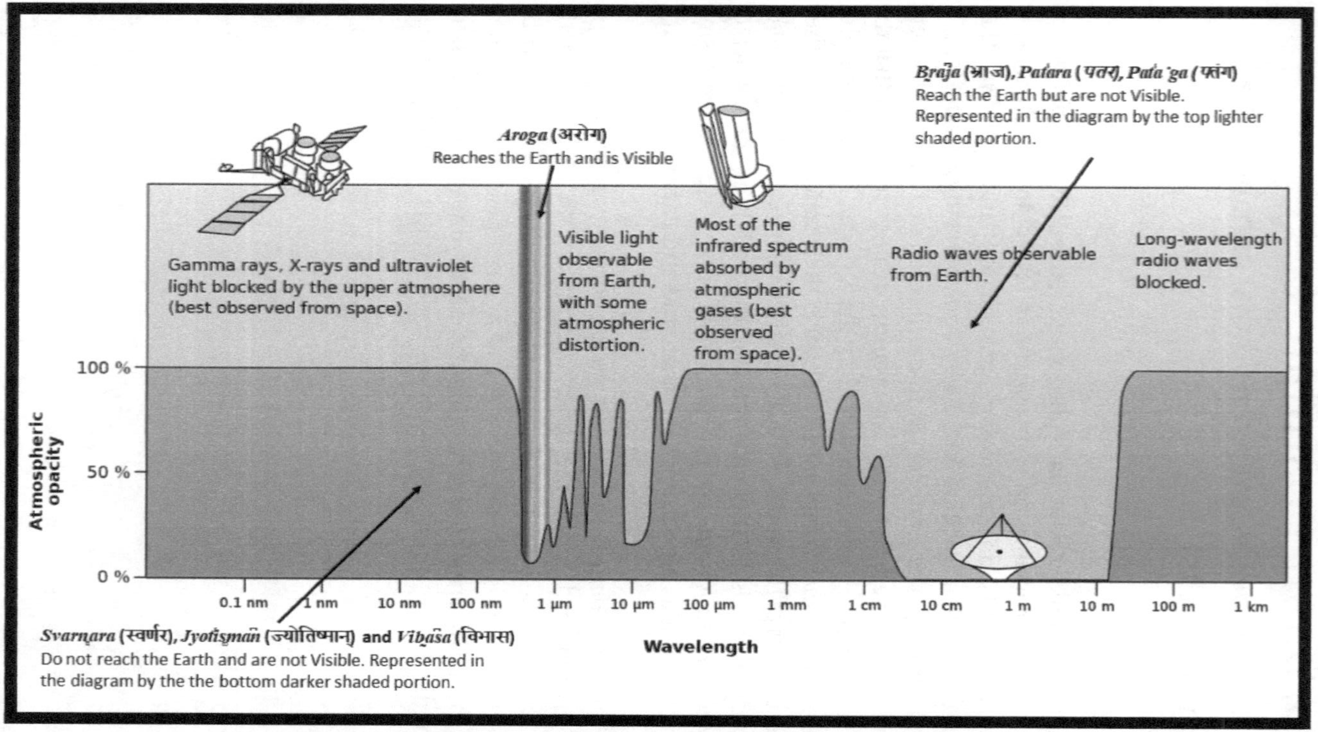

Aruṇapraśna's EMR labels superimposed on an existing NASA image.

Electromagnetic transmittance, or opacity, of the Earth's atmosphere, NASA (original); SVG by Mysid., Public domain, via Wikimedia Commons.

Sūryaraṣmi Postulate 6.3

The classification of the *Saptaśūryas* into ones that shine and ones that don't in *Aruṇapraṣna* 1.7.8 is a metaphoric division of the electromagnetic spectrum into radiations that reach the earth's surface and the ones that are absorbed or reflected by the earth's various atmospheric layers.

Aruṇapraṣna 1.7.8 describes the location of the *Saptaśūryas* in relation to the path to *Mahāmeru.* In this context, it is interesting to note that *Aruṇapraṣna* identifies the *Āroga* as the center of the radiation spectrum and defines it to be commonly seen visible light. It continues that along with *Āroga*, *Ḇrāja*, *Paśara* and *Paśa˙ga Sūrya* reach the earth's surface because they are below the path to *Mahāmeru. Svarnaro*, *Jyotiṣmān* and *Viḇāsa Sūryas*, on the other hand, are situated above the path to *Mahāmeru* and do not reach earth's surface. While a lot more research is needed to identify which of these six correspond to what components of the EMR spectrum, we can definitely say that *Ḇrāja*, *Paśara* and *Paśa˙ga* are the parts of the EMR spectrum that reach the earth's surface through the atmosphere. Likewise, the *Svarṇara*, *Jyotiṣmān* and *Viḇāsa* are absorbed or reflected away in the upper reaches of earth's atmosphere. We can also infer clearly that other than *Āroga, the other* six *Sūryas are not visible to the human eye.*

Conclusion

Solar radiation is the earth's principal source of energy and is understood to be a spectrum of radiations broadly divided into seven groups. Four of the radiations reach the earth's surface and the remaining three are filtered or reflected away by the atmosphere. These radiations are metaphorically represented by the *Saptaśūryas* in *Aruṇapraṣna.* Following today's science, *Aruṇapraṣna* also divides the *Saptaśūryas* into two groups, four that reach the earth's surface and three that do not. More research is needed to understand the one-on-one correlations between the *Sūryas* and the radiations of the solar electromagnetic spectrum. *Aruṇapraṣna*'s discussion on *Soma Devatā* enhancing the *Saptaśūryas* by using the eighth, *Sūrya Kṣyapa,* who only shines on *Mahāmeru* is a metaphoric representation to some form of a force field and an energy transfer process, the details of which need more research.

The potential areas for further research based on this topic include:

- Correlation between the components of the electromagnetic spectrum and the individual *Saptaśūryas.*
- Relationship between Seasons and Electromagnetic Radiation.
- Comprehensive study of *Saptaśūryas* as descriptions of electromagnetic radiation in the *Vedic* and *Puranic* corpus.
- Developing a scientific understanding of *Soma Devatā* in the context of electromagnetic radiation.
- Developing a scientific understanding of *Kṣyapa* in the context of electromagnetic radiation.

Chapter 7
Yajṉ̈enabanḓu

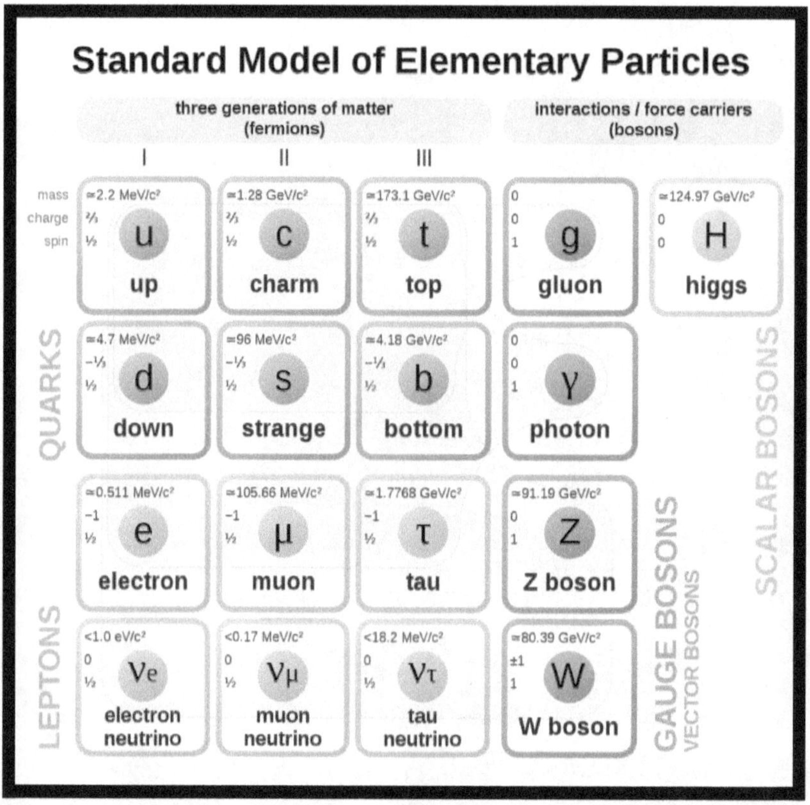

Cush, Standard model of elementary particles: the 12 fundamental fermions and 5 fundamental bosons.

Cush, Public domain, via Wikimedia Commons.

Science

The reality which we experience in our everyday lives is based on our interactions with our surroundings and is grounded in certain beliefs. If you are taking an evening stroll on a beach, it is impossible for you to be physically sitting at home and listening to music at the same time. The sand on the beach is solid material and the sound of the waves crashing on the beach is not. Whether you look at the other people walking on the beach or not, they are still there. The toy airplane that children are playing with does not magically move about in the air. It is controlled by signals from the handheld remote. If you parked your vehicle in a handicapped spot at the beach today, then you can get a citation for that from the officer, not because you are going to park incorrectly next month. Our experiences which agree with our intuition, like these beach experiences, may appear as mysteries on the surface but can be elegantly brought into the domain of our rational reality using classical Newtonian physics.

However, there is a counterintuitive reality we encounter when we explore the world of atoms and other subatomic particles where classical physics loses its relevance. In that world, objects can be in multiple places at the same time, can simultaneously behave like both solid particles and waves, things manifest only upon observation, objects can be correlated without any force mediating between them and cause and effect can become meaningless with effect sometimes preceding the cause.

Quantum physics is the science that delves into this mysterious world. It not only explains this world but also our everyday classical one as a special case of the quantum world. It has successfully explained the electromagnetic force, weak nuclear force, and strong nuclear force, which are the three of the four observed fundamental forces in nature. Quantum physicists are actively working on understanding the fourth force, gravity. In fact, scientists are continuously attempting to bridge the gap between the quantum and classical worlds using Collapse Models. From nuclear power to the wonders of the electronic revolution, quantum physics has contributed significantly to our current way of life.

As per quantum physics, the world we experience is composed of certain elementary particles. For example, all electromagnetic radiation is composed of an elementary particle called a photon. These particles exhibit energy that resembles that of both solid matter and waves simultaneously. Their properties cannot be described in the way we talk about the properties of physical objects in classical physics. In the classical world, we can define an object's location within the three-dimensional Euclidian space at any particular point in time, but we cannot do the same in the quantum world. A particle's location and many other properties can only be described using a probability function called a wave function, which quantifies the probability of the property being exhibited at any given time. Until an attempt is made to measure the property, we do not know a singular value for it with absolute certainty. At the instant at which we measure the property, its associated wave function collapses and it assumes a value. To illustrate with an example, an electron's location within the atom at any given instant of time is only a guess and the probability wave function is our attempt to measure the accuracy of that guess. It is a probability until we try to locate the exact, actual electron. Once we locate the electron by somehow capturing it, at that instant its location becomes certain at the location of capture. This and several other counterintuitive properties of particles have spawned an entire field of study under quantum physics devoted to particles called particle physics. One such particle phenomenon studied and the main subject of our discussion is quantum entanglement.

Quantum entanglement is a physical phenomenon in which two or more particles become related such that the quantum states of the entangled particles cannot be described independently of the other particles in the entangled group, even when these particles are separated at great distances. A quantum state of a particle is defined by a set of properties such as position, momentum, spin, and polarization. In a quantum entangled system, one or more of these properties are very strongly correlated. So, if we know the value of a correlated property of one particle, we can deduce the same for the other particles in the entangled system. Such an entangled group of particles is also referred to as a strongly correlated system. For particles in an entangled system, if we collapse the wave function of any particle in the group, the wave functions of the remaining particles simultaneously collapse. The affected property of each particle acquires a value according to the correlation relationship of the entangled system. Another related fact is that the probability distributions needed to define the quantum states of these particles deviate from the bell-shaped 'Gaussian' curves which underlie many natural processes.

Correlated systems are not exclusive to the quantum world, but it is unique because the correlation exists even when the entangled particles are separated from each other by great distances. We do not know what effectuates this correlation or if there is any physical transmission from one particle to the other, but we do know this change travels at speeds faster than speed of light! While physicists have been able to create and study quantum entanglement in a controlled environment and confirm the phenomenon exists, they are yet to uncover mother nature's secret behind it.

Quantum entanglement was first introduced by Albert Einstein, Boris Podolsky and Nathan Rosen as a thought experiment in a 1935 paper titled, "Can Quantum-Mechanical Description of Physical Reality be Considered Complete?[65]". The paper's objective was to show quantum physics as an inadequate model of reality. Their logic was that if quantum mechanics was a model that fully described reality, then it would be possible to conduct a measurement on one part of a quantum entangled system and instantaneously affect our knowledge about future measurements on the other part. This would mean that information can flow faster than the speed of light which is impossible with all known physics. They postulated that there are certain unknown agents or hidden variables that effectuate entanglement and remove dependence on information transfer between entangled particles at speeds faster than light.

The father of Quantum Physics, Neils Bohr argued against this Einsteinian view. His theories on this subject formed the roots of modern understanding of Physics and laid the foundation for the development of Quantum Physics we see today. Later in 1964 groundbreaking paper, "On the Einstein Podolsky Rosen paradox[63]," John Stewart Bell conceptually proved that the reality observed with quantum entanglement experiments does not agree with the existence of local hidden variables. He calculated a theoretical limit beyond which such correlations must have a quantum, rather than a classical, explanation. Since then, numerous experiments have confirmed the phenomenon of quantum entanglement but all of them have had one or more of several loopholes including problems with experimental design or set-up that affect the validity of the findings. In 2015, a group led by Ronald Hanson at the Kavli Institute of Nanoscience Delft in the Netherlands conducted a successful experiment without two of the major loopholes, detection efficiency and locality[69]. In addition, this experiment devised a better setup for producing entanglement on demand, giving confidence to engineers attempting to harness the power of quantum entanglement.

Entanglement is ubiquitous, so finding isolated, non-entangled particles is rare. In fact, the reality is a complex grid of entanglements. For example, electrons within any electron cloud are heavily entangled. Consequently, other than hydrogen, due to its singular electron, every other atom experiences entanglement in action. In fact, the helium atom is one of the most studied entangled systems due to its stability and inertness. Quantum entanglement has been demonstrated experimentally with photons, neutrinos, electrons, atoms, molecules as large as 60 carbon atoms and even with small diamonds. In the last few years, scientists have hypothesized that living systems use quantum entanglement in biological processes such as photosynthesis in plants and magnetic orientation in migrating species[68, 80]. In October 2018, physicists reported that they created quantum entanglement using living organisms, specifically, entanglement between living bacteria and quantized light[75]. Scientists, in another experiment, used green Sulphur bacteria to create entanglement between otherwise non-interacting light modes[71]. During the same year, three groups of physicists[67, 73, 74] went beyond producing entanglement between two particles or atoms and produced a large cloud of entangled atoms. They then split that cloud into groups, and found that they were still able preserve the quantum connection between the atoms inside.

Quantum entanglement is an accepted physical phenomenon, despite scientists not fully understanding how exactly it happens. The physical reality we experience manifests out of certain elementary particles which mostly exist in an entangled state rather than as free particles. There are some interesting aspects of reality that we can deduce based on this phenomenon. First, as seen in the living systems entanglement in the earlier paragraph, entanglement can happen between two vastly different types and sizes of entities. To elaborate through a thought experiment, say two particles, Photon A and Photon B, get entangled and subsequently get separated. Say the separated Photon A gets absorbed into an oxygen atom while Photon B is safely stored in a container in a lab. Neither the separation of the photons nor the absorption of the separated photons into other entities undoes the entanglement. Instead, the entire oxygen atom and Photon B will now be entangled. If this oxygen atom becomes part of a water molecule, then the water molecule also becomes entangled with Photon B. If the water molecule is mixed up in a body of water, then that water body as a system becomes entangled with Photon B. If the water dries up, becomes water vapor and eventually ends up in a cloud, then that cloud as a system is now entangled with Photon B. If instead of staying in the lab, Photon B somehow gets absorbed into the body of one of the scientists working at the lab, then the scientist and the cloud become entangled!

A discussion on quantum entanglement is not complete without talking about quantum decoherence, the loss of quantum entanglement behavior similar to energy loss from friction in a classical physics system. Laboratory created entangled systems are stable only as long as they are isolated. These man-made entanglements decohere in a fraction of nanosecond once they are allowed to interact with the outside environment, including our attempts to observe the phenomenon. There is no real loss of entanglement due to an interaction between a previously isolated entangled system and the environment. There is only an apparent loss. This is because when an isolated entangled system is allowed to interact with the environment, the entangled particles of the isolated system absorb into bigger and more complex systems in the environment, as explained in the clouds example in the earlier paragraph. This creates more complex entangled systems. Successive, uncontrolled entanglements ensue. In the midst of all this additional complexity, the originally isolated entanglement becomes unobservable. This loss in the ability to decipher the isolated entanglement is called quantum decoherence. Decoherence is one of the major limiting factors in utilizing quantum entanglement for practical applications such as quantum computing. It is interesting to note that entanglements that occur in certain natural systems seem to sustain for significantly longer periods of time compared to the entanglements created in the laboratory. For example, it has been observed that certain migratory birds that use entanglement to sense subtle changes in the Earth's magnetic field are able to sustain their entanglements for tens of microseconds[68]. Substantial research is needed to identify similar or longer lasting entanglements in nature.

Our understanding of quantum entanglement is far from complete. In fact, we do not even know why it happens. However, that does not stop us from attempting to utilize it for our benefit. Universities, corporations and various private and public research organizations around the world are trying to harness the power of quantum entanglement. These include quantum computing, quantum cryptography and quantum teleportation.

While a discussion into these applications is beyond the scope of this book, there is a point to be emphasized here: we do not need to know the underlying physics of a certain phenomenon to utilize it. The migratory birds that use quantum entanglement and the plants that could be using the same for photosynthesis do not study advanced physics and understand it before they use it. Therefore, it should not be a surprise if we humans have been utilizing this phenomenon for our benefit for thousands of years around the world without having studied quantum physics like scientists do today. Our ancestors may have had their own methods of understanding this natural phenomenon and devised methods to utilize it for their benefit. My quantum entanglement postulates are based on this simple observation. In this chapter, we are discussing one such application of quantum entanglement and then extending the hypothesis to other related applications.

Yajnenabandu Postulates

Postulate 7.1
Yajna, the fire sacrifice as outlined by *Veda* may be a quantum entanglement engine.

Postulate 7.2
The *Arunaketukagni Yajna*, which is the main purpose of *Arunaprasna* may be utilizing the entanglement amongst water bodies across the universe.

Postulate 7.3
Temple rituals, *Yoga* and all other non-*Yajna* practices in Hinduism and various rituals and practices in the different ethnic religions around the world may be utilizing quantum entanglement for the benefit of their practitioners.

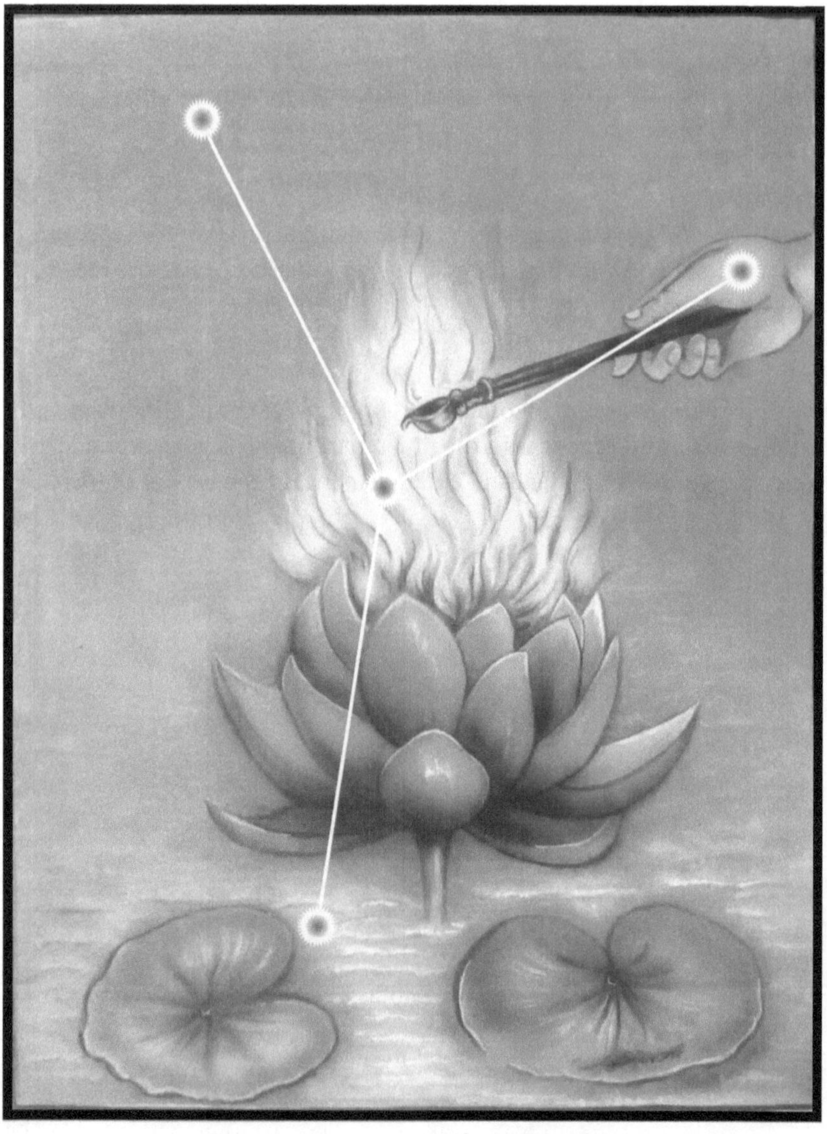

Yajña is a Quantum Entanglement Engine.

Yajṇenabaṇḍu Postulate 7.1

Yajṇa, A Quantum Entanglement Engine.

Yajṇa, the ritual fire sacrifice, is at the core of the *Vedic* way of life. From the outside, it seems like a simple act in which certain oblations are offered to the chosen *Devatā* through the sacred fire. However, there is a lot more to it than meets the eye. The *Yajṇa*s we talk about here are the ones that are done for the *Ṣrota* and *Gṛhyakarma* (गृह्यकर्म) of the *Kalpasūtras*(कल्पसूत्राः). The *Kalpasūtras*, which are succinct manuals that outline the process for conducting these *Yajṇa*s, classify the rituals into the *Nitya* (नित्य), *Nemittika* (नैमित्तिक), and *Kāmya* (काम्य) categories. The first two are obligatory and constitute a system of rituals that follow the cycles of the sun and the moon, or the stages of the human life from conception to death. The various obligatory rituals, though conducted independent of each other, are related to one another in the sense that they follow a strict sequence. The sequence and the dependence in the obligatory *Ṣodaṣakarmas* (षोडशकर्मानि) for birth, marriage, death etcetera as per *Gṛhyasūtras* (गृह्यसूत्राः) is self-explanatory. There is a similar pattern for the obligatory *Ṣrota Yajṇa*s as well but the sequence is based on layering complexity. The later and more complicated rituals in the *Ṣrota* sequence cannot be conducted unless the earlier and relatively simpler ones are done for a certain duration. For example, *Nityāgnihotra* (नित्याग्निहोत्र), the morning and evening fire ritual to *Sūrya* and *Agni* respectively has to be done for one year before one becomes qualified to perform the new moon and full moon obligatory rites, *Darṣapūrṇamāsiṣṭi* (दर्शपूर्णमासिष्टि). It is important to understand that each successive *Ṣrota* ritual does not replace the lower ones, but is layered on as an additional cyclical ritual. So, in our example, after the first year, the performer continues to perform the *Nityāgnihotra* every day. Now, in addition to that ritual, he performs the *Darṣapūrṇamāsiṣṭi* during every new moon and full moon. Similarly, each new layer of the system is gradually added as the performer gains eligibility to perform successive rites, from *Nityāgnihotra* to *Somayāga* (सोमयाग). Obviously, the *Ṣodaṣakarmas*, or the rituals that follow the stages of one's life from birth to death, do not follow this layering and only have a sequence. The *Ṣodaṣakarma* rituals, *Ṣrota* rituals and all other obligatory rituals are related to each other. The efficacy of one obligatory ritual depends on the efficacy of all the other rituals.

Not only are the obligatory rituals related to each other, but also they are related to the person for whom the ritual is conducted. The priests conducting the ritual and the entire ritual setup including the sacred fires, implements, and oblations are all an integrated system. My hypothesis is that this integrated system is an engineered quantum entanglement engine. This engine is developed in steps by the successive layers of constructed quantum entanglements that are established as the performer progresses through the sequence of rituals.

A lot more research is needed to identify and explain the various entanglements within this complex system, but for now we can hypothesize that there are two primary interconnected entanglements that are established and sustained through this fire sacrifice quantum engine. The first one is between the performer and the *Yajna* fires and the second between the *Yajna* fires and the macrocosm. The first entanglement is established through an elaborate preparatory ritual called *Agnyadana* (अग्न्याधान), which is conducted by priests who have been performing the entire sequence of rituals over a number of years. During this preparatory ritual, three to five sacred fires are established by the performer. The customized entanglement between these fires and the performer forms the foundation of this system. Therefore, these initial fires are kept alive until the death of the performer.

The fire is produced through friction using an apparatus called *Arani* (अरणि), which is built from wood of very specific trees, *Asvatta* (अश्वत्थ) (Ficus religiosa) and *Sami* (शमि) (Prosopis specigera). The science here could be that the dormant wood of the tree contains particles that are entangled to the greater macrocosm, and when the fire is produced from this wood these particles are absorbed into the performer's body, integrating the performer with the macrocosm. Alternatively, it could be that the fire produced from this *Arani* made of specific woods generates certain unique entangled particles which get separated. Some of these particles are absorbed into the performer's body, and their entangled counterparts are absorbed into the greater macrocosm.

Until recently, it was believed that entanglement could only be maintained in an isolated environment under extremely cold conditions to prevent decoherence. It was recently found that the exact opposite was true in an experiment conducted at ICFO in Barcelona, Spain[73]. The entanglement not only sustained in hot and chaotic conditions, but it actually flourished under those conditions. The entanglements created by priests in a *Yajna* are susceptible to decoherence as well. That is why a great effort is required to sustain them and/or replenish the system depleted by natural decoherence. It is also why the fire rituals are performed every day before sunrise and after sunset. A point to note here is that the setup includes a husband-and-wife couple. To avoid the consequences of decoherence, the couple must live a very disciplined life tightly synced to the cycles of the sun and the moon. The Sutras say that departing from this discipline breaks the connection to the sacred fire and must be reestablished as if they are starting all over.

There is also a lot of unknowns regarding how the *Yajna*'s fire is entangled with the greater macrocosm. One reason could be that since all creation originated from a single source, there is entanglement amongst its various parts. Almost all ethnic religions believe in the interconnectedness of all creation, and in the last few decades, numerous works have been published on this subject. My contention here is that *Yajna* uses this natural entanglement in some capacity to make a connection between the *Yajna* fire and the greater macrocosm. We must remember that successive absorption of entangled particles within larger bodies entangles those bodies as well. Advances in this line of research will help shed light on this macrocosmic entanglement.

A lot of science and engineering is involved in designing and building a quantum entanglement setup in a constrained laboratory environment. One may find it unbelievable that a fire generated by friction between pieces of wood is able to produce this complex phenomenon in an open environment. Science is study of nature. Engineering utilizes science for human needs. The various phenomena of mother nature happen whether or not we explain it scientifically. Moreover, humans have utilized various natural phenomenon for their benefit without understanding it the same way as scientists do. Let us look at some examples to illustrate this point. The human brain, a magnificent work of nature, has been behind all human accomplishments for thousands of years. Humans have utilized its capabilities to accomplish exceptional things in all walks of life without any schooling, leave alone advanced research in cognitive science. *Ganapatis* (घनपाठी) of the *Rgveda* memorize the singing of 10,600 of its verses along with all its pitch and tonal complexity in 11 different ways, and kept this tradition alive for thousands of years. The Romans, Egyptians, Chinese, Aztecs, Cambodians and Mayans did not have any computers or advanced materials, but still built massive cities and structures, some of which cannot be replicated even today with all our technology. Indigenous Hawaiians, just a few years ago, demonstrated that they can navigate the oceans without any gadgets, not even a compass. Every day, scientists are discovering new natural phenomenon where entanglement is utilized and are beginning to accept that it is as common as photosynthesis, which itself is possibly one of the miracles of entanglement.

Collaborative research between scientists and practitioners of *Srota Yajnas* will lead to greater insight into how this is accomplished in a *Yajna*. This can have tremendous benefit for engineers struggling to sustain generated entanglement from the natural chaos of decoherence. The research into the quantum entanglements of *Yajna* is difficult because the setup has a far more complicated space-time matrix. The integrated system of *Yajna* includes not only inanimate objects of the *Yajna* but also humans, the performing husband and wife couple and the priests and continues over many years in sync with space and time. It is a great area of multidisciplinary research including quantum physics, human anatomy & physiology, and cognitive science.

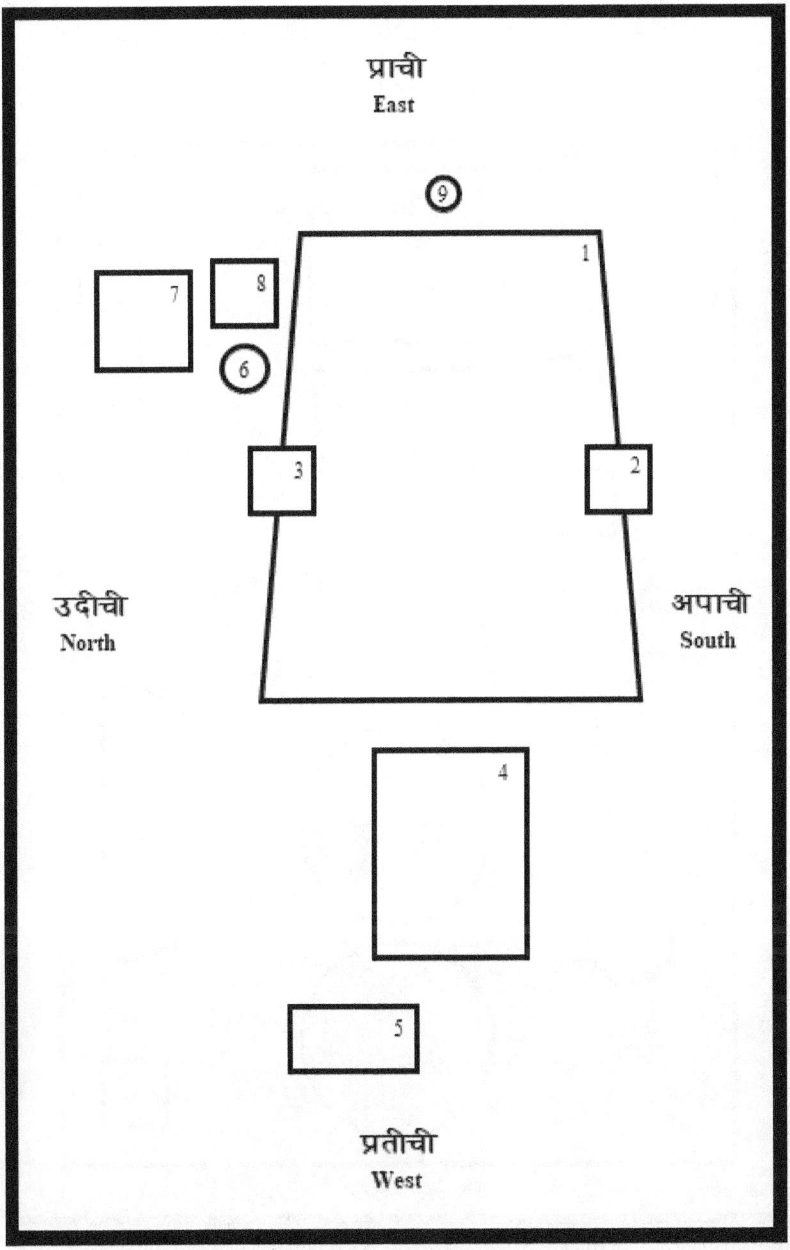

Approximate layout of a *Ṣrōta Vihāra.* Not to scale. Please refer to the legend provided at the end of these series of images.

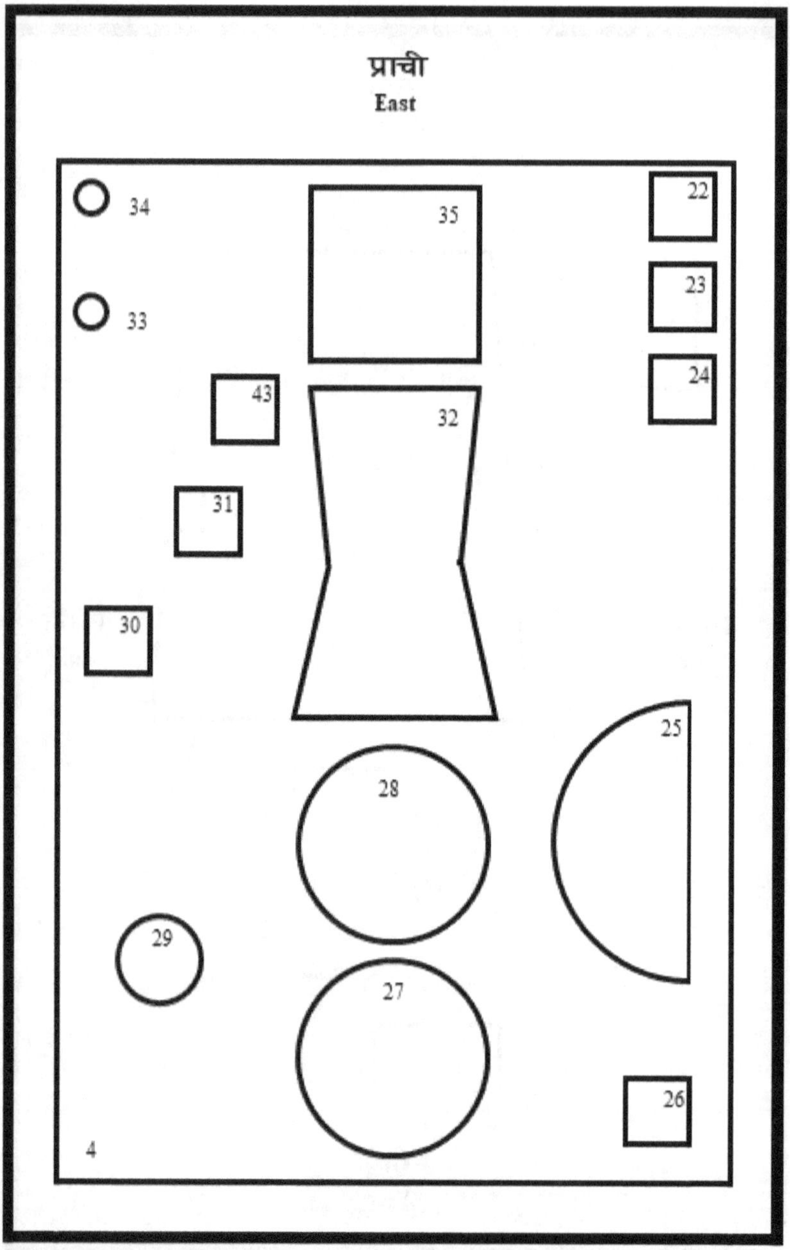

Approximate layout of the *Pracīnāva ṡa* section (Rectangle labeled number 4) of the *Ṣroṫa Vihāra.* Not to scale. Please refer to the legend provided at the end of these series of images.

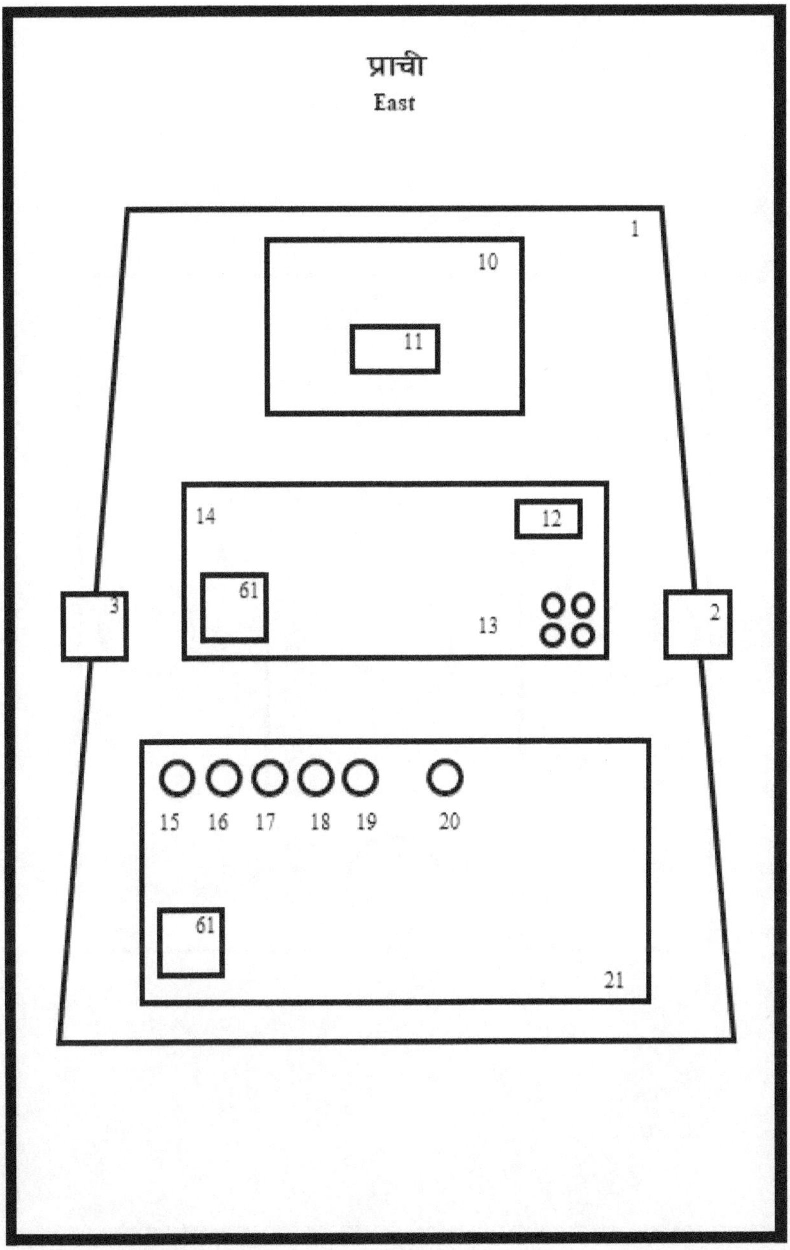

Approximate layout of the *Mahāvedi* section (Trapezium labeled number 1) of the *Śrota Vihāra.* Not to scale. Please refer to the legend provided at the end of these series of images.

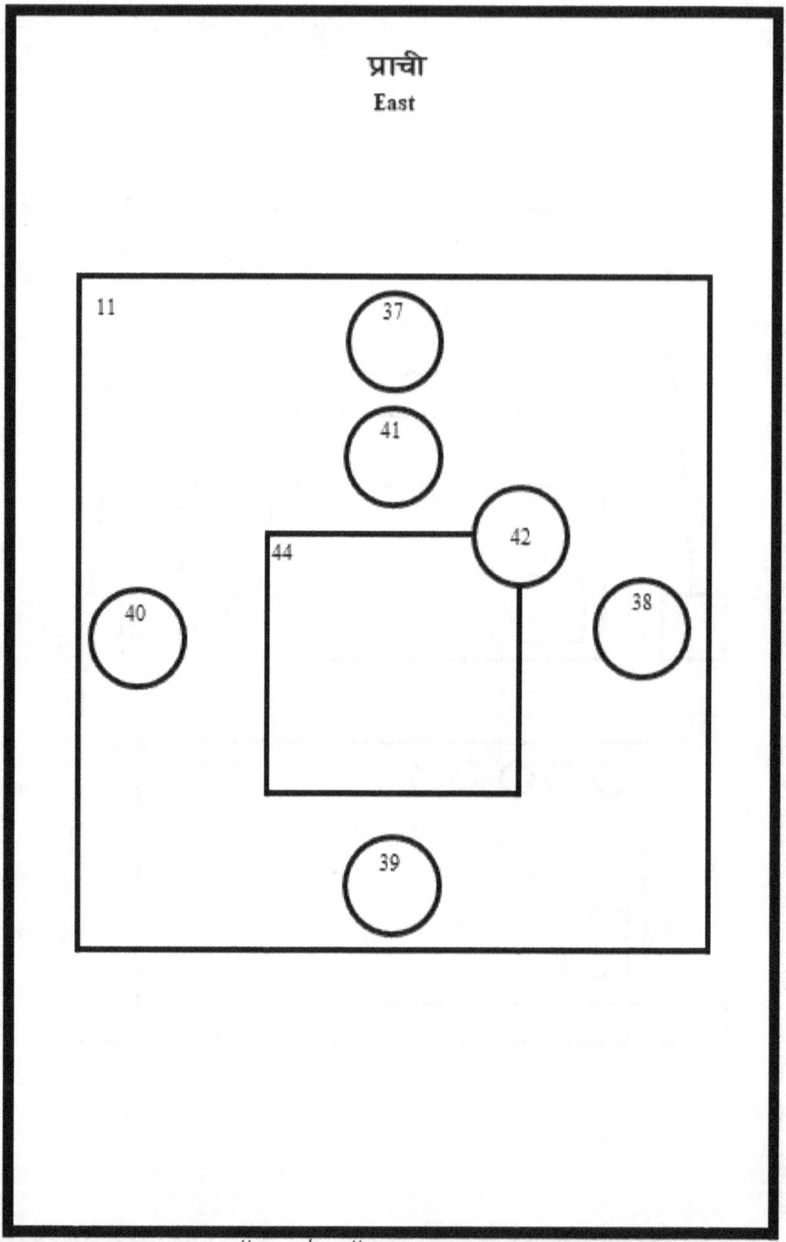

Approximate layout of the *Uttaravedi Citti* section (Rectangle labeled number 11) of the *Mahāvedi*. Not to scale. Please refer to the legend provided at the end of these series of images.

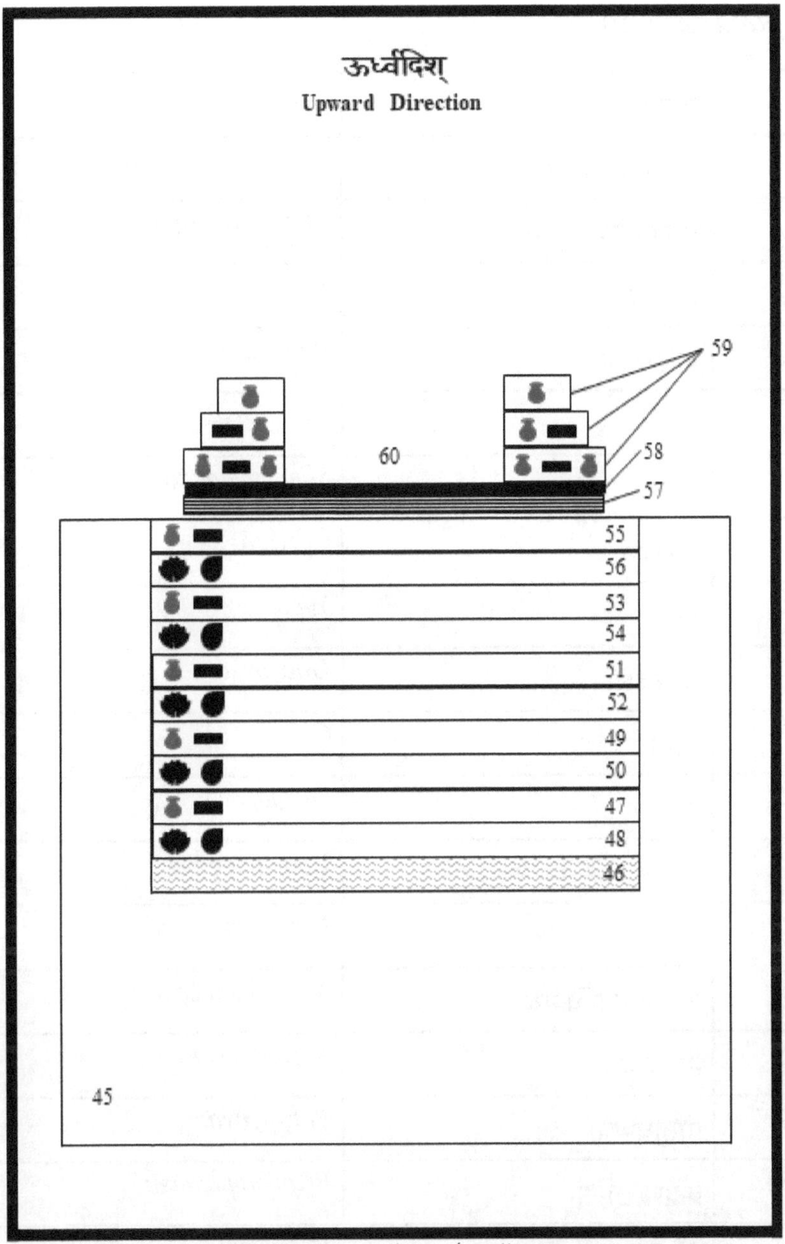

Approximate cross-sectional layout of the *Uttaravedi Citti* section (Rectangle labeled number 11) of the *Mahāvedi*. Not to scale. Please refer to the legend provided at the end of these series of images.

Vihāra Legend

1	महावेदि	*Mahāvedi*
2	आग्नीध्रीयमण्डपः	*Āgnīḍrīyamaṇdapa:*
3	मार्जालीयमण्डपः	*Mārjālīyamaṇdapa:*
4	प्राग्वंश	*Prāgva˙ṣa*
5	पत्निशाला	*Patniṣālā*
6	उतकरः २	*Utakara: 2*
7	शामित्रमण्डपः	*Ṣāmitramaṇdapa:*
8	चात्वालः	*Cātvāla:*
9	यूपः	*Yūpa:*
10	उत्तरवेदि	*Uttaravedi*
11	चितिः	*Citi:*
12	ग्रहखरः	*Grahaḵara:*
13	उपरव	*Uparava*
14	हविर्धानमण्डपः	*Havirḍānamaṇdapa:*
15	अच्छावाकधिष्ण्यः	*Acçāvākaḍiṣṇya:*
16	नेष्टधिष्ण्यः	*Neṣtaḍiṣṇya:*
17	पोतृधिष्ण्यः	*Potṛḍiṣṇya:*
18	ब्रह्मणाछन्सिः	*Brahmaṇāçansi:*
19	होतृधिष्ण्यः	*Hotṛḍiṣṇya:*

20	मैत्रावरुणधिष्ण्यः	*Mētrāvaruṇaḍiṣṇya:*
21	सदोमण्डपः	*Sadomaṇdapa:*
22	सदस्यः	*Sadasya:*
23	ब्रह्मः	*Brahma:*
24	यजमानः	*Yajamāna:*
25	दक्षिणाग्निः	*Ḍakṣiṇāgni:*
26	यजमानपत्नी	*Yajamānapatnī*
27	गार्हपत्याग्निः	*Gārhapatyāgni:*
28	अन्तरलयमा	*Antaralayamā*
29	प्रवर्ग्यखरः	*Pravargyakara:*
30	होतृः	*Hotṛ:*
31	अग्नीध्रः	*Agnīḍra:*
32	यजमानवेदि	*Yajamānavedi*
32	आह्वानीयाग्निः	*Āhvānīyāgni:*
33	प्रणेतः	*Praṇeta:*
34	उतकरः १	*Utakara: 1*
36	औदम्बरी	*Ōdumbarī*
37	वृष्टिजलघटः	*Vṛṣtijalagata:*
38	कूपजलघटः	*Kūpajalagata:*
39	सरोवरजलघटः	*Sarovarajalagata:*

40	नदीजलघटः	*Naḍījalagata:*
41	गृहजलघटः	*Gṛhajalagata:*
42	तडागजलघटः	*Taḍāgajalagata:*
43	अध्वर्युः	*Adhvaryu:*
44	अबीष्टकानिर्माणं	*Abīṣṭakānirmāṇa*
45	चित्याः लम्बरेखपरिच्छेदविस्तरः	*Cityā: Lambarekhaparic-chedavistara:*
46	भूमि	*Bhūmi*
47	अबीष्टकाः	*Abīṣṭakā:*
48	पुष्करकाण्डास्पर्णानिपुष्पानिच	*Puṣkarakāṇḍāsparṇānipuṣ*
49	अबीष्टकाः	*Abīṣṭakā:*
50	पुष्करकाण्डास्पर्णानिपुष्पानिच	*Puṣkarakāṇḍāsparṇānipuṣ*
51	अबीष्टकाः	*Abīṣṭakā:*
52	पुष्करकाण्डास्पर्णानिपुष्पानिच	*Puṣkarakāṇḍāsparṇānipuṣ*
53	अबीष्टकाः	*Abīṣṭakā:*
54	पुष्करकाण्डास्पर्णानिपुष्पानिच	*Puṣkarakāṇḍāsparṇānipuṣ*
55	अबीष्टकाः	*Abīṣṭakā:*
56	पुष्करकाण्डास्पर्णानिपुष्पानिच	*Puṣkarakāṇḍāsparṇānipuṣ*
57	काष्ठमञ्चः	*Kāṣṭamañca:*
58	गोमयं	*Gomaya*
59	अबीष्टकाः	*Abīṣṭakā:*

60	अग्निः	*Agni:*
61	उद्गातः	*Udgāta:*

Yajṇenabaṇḍu Postulate 7.2

Aruṇaketukāgni Yajṇa - A Quantum Entanglement amongst Water Bodies

The *Aruṇaketukāgni Yajṇa* is a *Kāmya* ritual, a non-obligatory *Ṣrota* ritual conducted for the purpose of seeking a specified benefit for the performer. *Aruṇapraṣna*'s entire purpose is to perform this ritual. *Kāmya* rituals can be done only by one currently performing the entire cycle of obligatory rituals. There must be certain physical reasons for this constraint. My interpretation of this is that the entanglement achieved and sustained through the obligatory ritual cycles provides the necessary foundation needed for the *Kāmya* ritual. This allows the layering of further entanglements between the microcosm and macrocosm, and the utilization of these connections for securing the benefits for the performer.

However, there is something very interesting about the *Aruṇaketukāgni* ritual: the altar for the sacred fire is built within a body of water. Water plays the central role in the cosmology and cosmogony of *Aruṇapraṣna*, which states that water is present everywhere in the universe. The 22nd *Anuvākā* of *Aruṇapraṣna*, the well-known *Mantrapuṣpam* (मन्त्रपुश्पम्), presents this concept very clearly. In the Bhoomi chapter of this book, we discussed *Aruṇapraṣna*'s cosmological model, which explains that the entire universe manifested out of primordial waters. The *Aruṇaketuka* ritual is in one way a symbolic microcosmic recreation of the universe. My second postulate for this chapter is based on water as well. I postulate that, *Aruṇaketukāgni* ritual involves building a quantum entanglement engine that builds and utilizes the entanglements between bodies of water.

Aruṇaketuka is a *Cayana* (चयन) type *Yajṇa* and so the sacrificial complex or *Vihāra* (विहार) that is erected for it is similar to that of other *Cayana* sacrifices, especially the *Sāvitra Cayana* (सावित्र चयन). *Sāyaṇācārya*, in his *Ḅāṣya* on *Aruṇapraṣna* in the very first *Mantra* of the first *Anuvākā*, *Aruṇapraṣna* 1.1.1 says that the *Sāvitra Cayana* is to be remembered for the *Aruṇaketukāgni* ritual. Again, in his *Ḅāṣya* on *Aruṇapraṣna* 1.22.12, he says that the procedure for erecting the *Aruṇaketukāgni* altar begins with drawing on the ground the *Raṭacakra* (रथचक्रं) of the *Sāvitra Cayana*. In the *Boḍāyana Śrotasūtra* (बौधायन श्रौतसूत्र), the procedure for the *Aruṇaketuka* is described in the 10th *Anuvākā* of the 19th *Praṣna* and it ends with the instruction that subsequent procedure after what has been described there follows that of *Sāvitra Cayana*. The full details of the *Aruṇaketukāgni* ritual setup and procedure are beyond the scope of this introductory work and will be the subject matter of a subsequent publication. For the purposes of our discussion here, a brief description of the same would suffice.

The *Vihāra* consists of two main sections, the *Prācīnāva ṣa* (प्राचीनावंश) and the *Mahāvedi* (महावेदि). The former has the setup which is mostly similar to the ones for the daily fire ritual and the new moon and full moon rituals and has the *Tretāgni* (त्रेताग्नि), three main fundamental altars, namely *Gārhapatya* (गार्हपत्य), *Dakṣiṇāgni* (दक्षिणाग्नि) and *Āhvanīya* (आह्वानीय) among other related structures. The *Mahāvedi* consists of 3 main subsections, *Sadasa* (सदस), *Havirdāna* (हविर्धान) and *Uttaravedi* (उत्तरवेदि). The layout of the *Vihāra* is mainly on the east-west axis with the *Uttaravedi* being the easternmost section. It is in this area of *Uttaravedi* that a specially designed altar is erected for *Aruṇaketukāgni* ritual.

The altar consists of a complex arrangement of various layers of specialized bricks, which is only superficially described in *Aruṇapraṣna*. A better description of these various bricks is presented in the *Boḍāyana Śrotasūtra* 19.10 but even that is not detailed enough either and to get a complete understanding one must rely on the practitioners only. That being said, for our purposes of explicating my water quantum entanglement postulate, the description in the *Śrota Sūtras* along with the *Sāyaṇācārya Bāṣya* on *Aruṇapraṣna* is sufficient.

First the *Raṭacakra* consisting of nine concentric circles are drawn on the ground. This step is similar to *Sāvitra Cayana*. Then a knee-deep pit as measured against the performers knee is dug and that pit is filled with water up to the depth to submerge the performer's ankle. A piece of gold, an idol of a human and tortoise is placed in these waters. The water surface in the pit is then covered with lotus flowers, lotus leaves, lotus stems, and lotus roots. On top of the lotus vegetation, an altar made of *Abīṣṭakā:* (अबीष्टका:) is built. The exact definition of *Abīṣṭakā:* is not clear in the texts. In practice, these have included a combination of porous stones, earthen pots filled with water from various sources and other specialized materials known to the practitioners.

The lotus vegetation layer topped with *Abīṣṭakās* is considered as one layer or *Citti* (चित्ति). Five such layers are constructed one over the other. These five *Cittis* symbolize the five altars, the *Tretāgni* mentioned above and the two *Sabya:* (सभ्य:) and *Avāstya:* (अवास्थ्य:). On top of this a wooden log platform is erected. On this platform a layer of specially sourced cow dung is spread. It is on this that a fire altar for the ritual is established. This sacred fire that has been set up is then secured on all four sides by *Abīṣṭakās.*

Around this fire altar, water that has been sourced from five different sources and poured into earthen pots is placed. Rain water that was collected in a vessel is placed to the eastern end of this pit. Water from a well is placed to the south of the same. Water from a deep and non-flowing lake is taken and placed on the west. Water from a flowing river is placed on the north. Water that has been stored in pots at home is brought and put between the pots of flowing river water in the north and the fire altar but symbolically considered to have been placed in the middle. Water from shallow ponds and lakes is brought and placed on top of the *Abīṣṭakās* surrounding the sacred fire. This is more or less the core setup of this ritual as it pertains to our discussion. Numerous oblations and related activities are done by the performer and his wife with the guidance and help from at the least sixteen priests.

Since water is seen everywhere and used every day, we implicitly assume that scientists who have made so many advances in understanding the secrets of nature, know everything that there is to know about water. Nothing could be farther than the truth. Water is still a great mystery for the scientists especially at a quantum level. In 2016, scientists at the Oak Ridge National Laboratory discovered that water molecules are capable of quantum tunneling[72]. A research team at University of

Houston found that quantum behavior of protons in tiny quantities of water are markedly different than those in bulk water[80]. In fact, they concluded that, quantum mechanics of protons in water has been playing a role in the development of cellular life all along but we never noticed before. Human body is more than 60% water. Every cell of the human body has water and these are tiny quantities of water and so the protons in these minute water bodies are in the quantum mechanical space.

Aruṇapraṣna 1.26.13 says that this *Aruṇaketukāgni* is kind of fire that is inside water. This could be referring to quantum energy fields within water. The *Aruṇaketukāgni Yajña* then is a process by which the water within the cells of the performer is entangled with the water from sources that is placed around the *Cittis*. The energy required for this entanglement is probably sourced from the prior established entanglements between the performer and the macrocosm through the obligatory sacrifices and channeled using the fires in this sacrifice. The entangled water in various pots then evaporates from the heat of the sacrifice and is absorbed into the atmosphere and through successive absorption or further entanglements gets connected with various water bodies in the greater macrocosm. I am not concluding on any defined process here as such but giving one among many possibilities. I am however postulating that there is quantum entanglement between the water bodies within the performers body with the water bodies in the greater macrocosm. This is a fascinating area for further research.

That there is a connection between the performer and water is clearly outlined in *Aruṇapraṣna* 1.26.12 - *Aruṇapraṣna* 1.26.15. These are given in form of *Niyamas* (नियमाः) or rules of conduct for one who has done this *Yajña*. Given the lengthy obligatory rituals sequence, it takes more than five years in one's life at a minimum and probably on the average ten years for one to be able to successfully complete this *Aruṇaketukāgni* sacrifice. Not only time, this is a very expensive affair as well for the performer. Given the time and resources involved, one is advised to follow these rules of conduct to prevent losing the connection established through this sacrifice. The rules include the following. One should not run when caught in rain. One should not spit, urinate or defecate in water. One should not sit on gold or lotus leaves. One should not eat any aquatic animals, especially tortoise. How doing these things leads to decoherence of established entanglement or in any way causes any harm to performers is another area of collaborative research.

The benefit from *Yajŋa* has long been a research topic. There have been many scholars and scientists who have attempted at establishing documented evidence of benefits of *Yajŋa.* However, they have barely scratched the surface of this treasure. My hypothesis is that one constantly interacts with the external environment to get what one wants. Entanglement of the performer with the greater macrocosm through *Yajŋa* enables the performer to make optimized decisions in the performer's interaction. The interaction consists of uncountable conscious and subconscious decisions one takes when engaging with the world. It is not possible to rationally analyze every decision based on millions of variables, most of which are not even palpable to the decision maker. This closed system enables the optimization of decision making and thereby secures the benefits for the performer. This is another area for multidisciplinary research.

Yajnenabandu Postulate 7.3

Practices within Hinduism other than *Yajna* and practices of other ethnic religions employ quantum entanglement as well.

Fire sacrifice or another form of sacrifice is not limited to Hinduism. Almost all ethnic religions including Shen, Roman, Greek, African, Australian Aboriginal and Native American cultures have in one way or another a sacrifice as an important aspect of their practices. Within Hinduism there are many other practices that do not involve the fire sacrifice including *Mūrti* (मूर्ति) worship, *Yoga* & meditation, personal rituals like *Sa'dyāva'dana* (संध्यावंदन). Likewise, with almost all ethnic cultures of the world there are rituals that do not use a fire sacrifice. My third postulate for this chapter is that all these practices within all ethnic religions of the world, ones that involve some sort of fire sacrifice and those that involve other forms of ritualistic practices involve some sort of quantum entanglement with various forces in mother nature for the benefit of the performer and these have been developed by the respective cultures as what suited best to their needs and surroundings. All these are fertile areas for entanglement based multidisciplinary research.

Conclusion

Quantum physics is the science that delves into the mysterious world of atoms and subatomic particles. An important quantum phenomenon is quantum entanglement. Our ancestors may have had their own methods of understanding this natural phenomenon and devised methods to utilize it for their benefit. *Yajṉa*, the fire sacrifice as outlined by *Veḏa* may be a quantum entanglement engine. The *Aruṉaketukāgni Yajṉa*, which is the main purpose of *Aruṉaprasṉa* may be utilizing the entanglement amongst water bodies across the universe. It is possible that various ethnic religions around the world may have such rituals and practices as well.

The potential areas for further research based on this topic include:

- Build a comprehensive database of theoretically possible quantum entanglements amongst participants, macrocosmic elements, ritual setup, and related objects in all the various *Sṟoṯa* rituals.
- Study of water from various sources as identified in *Aruṉaprasṉa* to understand the similarities and differences in their classical and quantum characteristics.
- Comparative study of quantum entanglement in a laboratory and the setup of a *Yajṉa* ritual.

अन्नाद्भवन्ति भूतानि पर्जन्यादन्नसम्भवः ।

यज्ञाद्भवति पर्जन्यो यज्ञः कर्मसमुद्भवः ॥ महाभारतम् ६-३-२७-१४ ॥

कर्म ब्रह्मोद्भवं विद्धि ब्रह्माक्षरसमुद्भवम् ।

तस्मात्सर्वगतं ब्रह्म नित्यं यज्ञे प्रतिष्ठितम् ॥ महाभारतम् ६-३-२७-१५ ॥

Glossary of Selected Sanskrit Terms

The transliteration of the *Vedic* words in the glossary are given with the case endings as applicable. In the text of the book and in the glossary meanings however the transliterated words are sometimes used without their case endings and sometimes the case endings are replaced with derived alternative endings for ease of reading in the English language. These include plural forms (*Purāṇas*), possessive forms (*Aruṇapraṣna's*) and derived adjectives (*Vedic*).

Devanāgarī	***Devanāgarī* Transliterated**	**Word Meaning**
अग्निः	*Agni:*	A *Devatā* associated with fire. *Agni* is central to all the *Vedic* rituals. Also, can be used to mean one of the five *Pañcabūtās*, namely the fire element.
अग्नेयाः	*Agneyā:*	A type of clouds that cause rain in upper worlds.
अग्न्याधानं	*Agnyādāna*	A *Śrota* ritual to initiate a *Gṛhasta* into the mandated practice of *Nityāgnihotra*.
अच्छावाकः	*Accāvāka:*	One of the three assistants of the *Hotā*.
अणुः	*Aṇu:*	Unit of time, approximately equal to 52.67 Microseconds.

Devanāgarī	*Devanāgarī* **Transliterated**	**Word Meaning**
अध्वर्युः	*Aḏvaryu:*	One who is responsible for all physical construction and ritual procedures in a *Yajṇa* including erecting the vihara, building various *Citis*, readying the sacrificial implements, gathering the wood, fetching the water, lighting the fire, and preparing the oblations.
अनादि	*Anādi*	Something that does not have a beginning.
अनुपलब्धि	*Anupalabḏi*	Knowledge through negative perception. One of the six *Pramāṇās*.
अनुमानं	*Anumāna*˙	Knowledge through reasoning or inference. One of the six *Pramāṇās*.
अनुवाकः	*Anuvāka:*	Section of a text.
अपूर्वं	*Apūrva*˙	Results of actions (*Vedic* rituals), which are physically not perceived at the time of the actions but produce physically perceivable results in the future.
अपौरुषेयः	*Aporuṣeya:*	Not authored by anyone.

Devanāgarī	*Devanāgarī* **Transliterated**	**Word Meaning**
अप्सरसः	*Apsarasa:*	A class of female *Devatās* associated with music and dance. Wifes of the *Gandarvās*.
अरणिः	*Araṇi:*	An apparatus made of wood of very specific trees, *Aśvatta* (Ficus Religiosa) and *Śami* (Prosopis Specigera) used to produce fire for *Yajña* using friction.
अरुणकेतुकः	*Aruṇaketuka:*	The central *Devatā* of the *Aruṇapraśna*.
अरुणकेतुकाग्निः	*Aruṇaketukāgni:*	See *Aruṇaketuka*.
अरुणप्रश्नः	*Aruṇapraśna:*	First *Prapāṭaka* of the *Tettirīya Āraṇyakam in the Kṛṣṇayajurveda.*
अरुणाः	*Aruṇā:*	A family of *Ṛṣis*.
आरोगः	*Āroga:*	One of the seven *Saptasūryās*.
अर्थशास्त्रं	*Artaśāstra*	An ancient treatise of *Bāratam* on statecraft, military strategy and economic policy.
अर्थापत्ति	*Artāpatti*	Knowledge through circumstantial implication. One of the six *Pramāṇās*.
अवास्थ्यः	*Avāstya:*	One of the five fires a *Śrota Gṛhasta* maintains.

Devanāgarī	*Devanāgarī* **Transliterated**	**Word Meaning**
अश्मिविद्विशः	*Aṣmividviṣa:*	One of the six *Vāyugaṇās*.
अश्वत्थः	*Aṣvaṭṭa:*	The sacred fig tree (Ficus Religiosa).
अश्विनौ	*Aṣvino*	Twin *Devatās* associated with health and medicine among other things.
असुराः	*Asurā:*	A class of superhuman beings.
अहङ्कारः	*Ahaṇkāra:*	Identification or attachment to one's ego.
आकाशः	*Ākāṣa:*	Space.
आग्नीध्रः	*Āgnīḍra:*	One of the three assistants of the *Brahma*.
आचार्यः अरविन्दः	*Ācārya: Aravinda:*	Sri Aurobindo.
आत्मः	*Ātma:*	The closest translation is one's core self. It is not one's body, mind or ego.
आदित्यः	*Āditya:*	A *Devatā* associated with the Sun.
आपः	*Āpa:*	One of the five *Paṇcabūtās*.
आयुर्वेदः	*Āyurveda:*	An ancient treatise of *Bāratam* on health, nutrition and medicine.

Devanāgarī	**Devanāgarī Transliterated**	**Word Meaning**
आरण्यकः	*Āraṇyaka:*	A section of the *Veda* dealing with both rituals and philosophical analysis.
आह्वानीयः	*Āhvānīya:*	One of the five fires a *Ṣrōta Gṛhasṭa* maintains.
इतिहासौ	*Itihāsō*	Historical narrative epics of *Ḅāratam*, *Rāmāyaṇam* and *Mahāḅāratam*.
इन्द्रः	*Indra:*	A *Devatā* associated with the thunder, lightning, rain and war among other things.
इन्द्रियं	*Indriya*˙	Sensory organs and their associated subtle attributes.
उत्तरवेदि	*Uttaravedi*	For Soma *Yajña* and other complex rituals, the eastern most section of the *Mahāvedi*, where located are the main *Citis* for the ritual.
उदकं	*Udaka*˙	Water.
उद्गातृः	*Udgātṛ:*	One who sings the *Vedic* hymns during a *Yajña*.
उपनिषद्	*Upaniṣad*	A section of the *Veda* dealing primarily with philosophical analysis.

Devanāgarī	***Devanāgarī* Transliterated**	**Word Meaning**
उपमानं	*Upamāna˙*	Knowledge through comparison and analogy. One of the six *Pramāṇās*.
ऋग्वेदः	*Ṛgveda:*	One of the five collections comprising the *Veda*. Each collection consists of *Sahitās*, *Brahmaṇas*, *Āraṇyakas* and *Upaniṣads*.
ऋतं	*Ṛta˙*	Regulating and coordinating natural order of the universe. The cycle of *Vedic* rituals keeps the human society in sync with this natural order. Its principles directly, and through the ritualistic lifestyle indirectly reflect in all aspects the *Vedic* society.
ऋतुः	*Ṛtu:*	Season.
ऋतुसंहारः	*Ṛtusa˙hāra:*	*Kālidāsa's* poetic work on seasons.
ऋषिः	*Ṛṣi:*	A seer of the *Vedic* tradition who has visualized the workings of *Ṛta* with their mind's eye and developed a life style for people of this tradition to live in sync with *Ṛta*.

Devanāgarī	*Devanāgarī* **Transliterated**	**Word Meaning**
कर्मकाण्डं	*Karmakāṇḍa˙*	The cycle of *Vedic* rituals that keeps the human society in sync with *Ṛta*.
कल्पसूत्राः	*Kalpasūtrā:*	One of the six *Vedāṅgas*. A collection of texts related to procedural details of the *Karmakāṇḍas* and social and moral code for the way of life in *Vedic* tradition.
काम्यकर्मन्	*Kāmyakarman*	Non-obligatory *Vedic* rituals that are conducted for securing specified benefits such as progeny, health, wealth, heaven etcetera.
कालिदासः	*Kālidāsa:*	Considered to be the greatest *Sa˙sakṛtam* poet. National poet of *Bāratam*.
कृष्णयजुर्वेदः	*Kṛṣṇayajurveda:*	One of the five collections comprising the *Veda*. Each collection consists of *Sahitās*, *Brahmaṇas*, *Āraṇyakas* and *Upaniṣads*.
केतवाः	*Ketavā:*	A family of *Ṛṣis*.
क्श्यपः	*Kṣyapa:*	A *Ṛṣi*.
गणाः	*Gaṇā:*	Group.

Devanāgarī	*Devanāgarī* **Transliterated**	**Word Meaning**
गन्धर्वाः	*Gand̰arvā:*	A class of male *Devatās* associated with music and dance. Husbands of *Apsarasas*.
गर्गः	*Garga:*	A *R̥ṣi*.
गार्हपत्यः	*Gārhapatya:*	One of the five fires a *Śrota Gr̥hast̰a* maintains.
गुरुत्वाकर्षण	*Gurutvākarṣaṇa*	Gravity.
गृहमेधः	*Gr̥hamed̰a:*	One of the six *Vāyugaṇās*.
गृहस्थः	*Gr̥hast̰a:*	A married householder.
गृह्यकर्मन्	*Gr̥hyakarman*	Obligatory *Vedic* rituals that are conducted at various stages of one's life from pre-conception to death.
गृह्यसूत्राः	*Gr̥hyasūtras*	One of the four sections of the *Kalpasūtrās*.
घनपाठीः	*G̰anapāt̰ī:*	The most advanced chanters of the *Veda*. *G̰anapāt̰īs* can recite in all the 11 ways (*pāt̰ās*) including the most difficult one, *G̰anpāt̰ā*.
घर्मः	*G̰arma:*	A prescribed quantity and mix of goat and cow milk.
चक्षुसौर्यः	*Cakṣusorya:*	One of the *Vedic R̥ṣis*.

Devanāgarī	*Devanāgarī* **Transliterated**	**Word Meaning**
चयनं	*Cayana*	A class of complex *Vedic* rituals.
चाणक्यः	*Cāṇakya:*	Ancient philopher, economist, jurist and royal advisor of *Bāratam*.
चातुरमास्यं	*Cāturamāsya*	The three seasonal *Nemittikakarmans*.
चितिः	*Citi:*	Fire altar of *Yajña*.
ज्योतिष्मान्	*Jyotiṣmān*	One of the seven *Saptasūryās*.
तन्मात्रं	*Tanmātra*	The essential elements that are the primordial causes of the *Pañcabūtās*.
तपस्या	*Tapasyā*	The spectrum of practices within the greater *Vedic* tradition that develop a progressive understanding of the self.
तैत्तिरीय आरण्यकम्	*Tettirīya Āraṇyakam*	One of the *Āraṇyakas* of the *Kṛṣṇayajurveda*.
त्रुटिः	*Truti:*	Unit of time, approximately equal to 0.30 Microseconds.
त्रेताग्नयः	*Tretāgnaya:*	*Gārhapatya*, *Āhvānīya*, and *Dakṣiṇāgni*.
त्वष्टः	*Tvaṣṭṛ:*	A *Devatā* associated with design, engineering and building.

Devanāgarī	***Devanāgarī*** **Transliterated**	**Word Meaning**
दक्षिणाग्निः	*Dakṣiṇāgni:*	One of the five fires a *Ṣrota Gṛhasṭa* maintains.
दर्शपूर्णमासिष्टिः	*Darṣapūrṇamāsiṣṭi:*	The *Nemittikakarmans* performed during the new moon and full moon days.
दृष्टान्तः	*Dṛṣṭānta:*	Empirical evidence.
देवः	*Deva:*	One of the three principal elements of *Yajṇa*. Within the greater *Vedic* tradition the word has a complex set of meanings and interpretations. It can be understood as the anthropomorphic representation of various aspects of *Ṛta*.
देवता	*Devatā*	See *Deva*:.
द्रव्यस्वभावः	*Dravyasvaḇāva:*	Properties.
द्विवचनं	*Dvivacana'*	Dual grammatical number. *Sasakṛtam* has three grammatical numbers, singular, dual, and plural.
द्योष्पितृः	*Dyoṣpitṛ:*	A *Devatā* associated with sky.

Devanāgarī	*Devanāgarī* **Transliterated**	**Word Meaning**
धर्मः	*Ḍarma:*	As per *Mīmāsakās*, *Ḍarma* is the right conduct as understood from correct intrepretation of the *Veda*. It seeks to keep the individual and society in sync with *Ṛta*. Within the greater *Vedic* tradition the word has a complex set of meanings and interpretations.
धूपयः	*Ḍūpaya:*	One of the six *Vāyugaṇās*.
धृव	*Ḍṛva*	Pole Star.
नामधेयः	*Nāmaḍeya:*	Name.
नित्यकर्मन्	*Nityakarman*	Obligatory *Vedic* rituals that are to be performed every day. The most important of which is *Nityāgnihotra*.
नित्याग्निहोत्रं	*Nityāgnihotra*	Daily obligatory *Vedic* ritual for *Sūrya Devata* just prior to sunrise and *Agni Devata* shortly after sunset.
निमेषं	*Nimeṣa*	Unit of time, approximately equal to 88.9 Milliseconds.

Devanāgarī	*Devanāgarī* **Transliterated**	**Word Meaning**
नियमं	*Niyama*	Rules of conduct for the performer and related indivudals before, during, and/or after a *Vedic* ritual. Within the greater *Vedic* tradition the word has been applied to similar contexts outside the ritual settings.
निरूढपशुबन्ध	*Nirūḍapaṣubanḍa*	One of the *Nemittikakarmans* performed during the rainy season.
नेष्टः	*Neṣta:*	One of the three assistants of the *Aḍvaryu.*
नैमित्तिककर्मन्	*Nemittikakarman*	Obligatory *Vedic* rituals that are to be performed as per cycles of nature such as *Pakṣas*, *Ṛtus*, and stages of one's life.
पक्षः	*Pakṣa:*	Lunar fortnight.
पश्चकर्णः	*Paṇcakarṇa:*	A *Ṛṣi.*
पश्चभूताः	*Paṇcaḇūtā:*	Five basic elements of the *Vedic* cosmology. Everything in the material world is a variegated combinations of these basic elements.

Devanāgarī	*Devanāgarī* **Transliterated**	**Word Meaning**
पतंगः	*Pata'ga:*	One of the seven *Saptasūryās*.
पतरः	*Patara:*	One of the seven *Saptasūryās*.
पद्धतयः	*Paddataya:*	Detailed guides to the *Srotasūtras*.
परमाणुः	*Paramāṇu:*	Smallest and indivisible particle of matter.
परमात्मः	*Paramātma:*	The closest translation is Absolute *Ātma*.
पर्जन्याः	*Parjanyā:*	Clouds that cause rain on Earth.
पश्यकः	*Paṣyaka:*	One who can see past, present, and the future.
पितराः	*Pitarā:*	Ancestors.
पितृः	*Pitṛ:*	Ancestors.
पिशाचाः	*Piṣācā:*	A class of beings who are flesh eating demons.
पुराणाः	*Purāṇā:*	Encyclopedic literature of *Bāratam* that covers diverse topics such as cosmogony, cosmology, genealogies of gods, goddesses, kings, heroes, sages, and demigods, folk tales, pilgrimages, temples, medicine, astronomy, grammar, mineralogy, humor, love stories, as well as theology and philosophy.

Devanāgarī	*Devanāgarī* **Transliterated**	**Word Meaning**
पूर्वमीमांसा	*Pūrvamīmā˙sā*	The exegesis of the *Veda* to understand *Ḍarma*.
पूशन्	*Pūṣan*	*Devatā* associated with marraiges, journeys and safety during journeys among other things.
पृथ्वी	*Pṛṭvī*	Earth and/or one of the five *Paṇcaḇūtās*, namely the earth element.
पोताः	*Potā:*	One of the three assistants of the *Brahma*.
प्रजापतिः	*Prajāpati:*	*Devatā* associated with creation among other things.
प्रतिज्ञा	*Pratijṇā*	Hypothesis.
प्रत्यक्षं	*Pratyakṣa˙*	Knowledge through sensory perception or extended sensory perception. One of the six *Pramāṇās*.
प्रपाठक	*Prapāṭaka*	Section of a text.
प्रमाणानि	*Pramāṇāni*	Means or sources of knowledge.
प्रयोगाः	*Prayogā:*	Detailed procedures for the rituals in the *Ṣrotasūtras*.

Devanāgarī	*Devanāgarī* Transliterated	Word Meaning
प्रवर्ग्यः	*Pravargya:*	A type of *Vedic* ritual, which is a part of the various rituals comprising the *Somayāga*.
प्राचीनावंशः	*Prācīnāva ̇ṣa:*	The western section of the *Vihāra*. It is mostly similar to the structures for the daily fire ritual and the new moon and full moon rituals and has the *Tretāgni* three main fundamental altars, namely *Gārhapatya*, *Ḍakṣiṇāgni*, and *Āhvānīya* among other related structures.
प्राणत्रातः	*Prāṇatrāta:*	A *Ṛṣi*.
प्रशास्ताः	*Prṣāstā:*	One of the three assistants of the *Hotā*.
प्लाक्षिः	*Plākṣi:*	A *Ṛṣi*.
बन्धु	*Banḏu*	Entanglement.
बाहुकः	*Bāhuka:*	A character from *Mahāḇāratam*.
बुद्धि	*Budḏi*	Intellect
बृहस्पतिः	*Bṛhaspati:*	A *Devatā* associated with light, knowledge, and *Ṛta*.
बौधायन श्रौतसूत्रम्	*Boḏāyana ṣrotasūtram*	One of the *Ṣrotasūtras*.

Ḋevanāgarī	*Ḋevanāgarī* **Transliterated**	**Word Meaning**
ब्रह्म	*Brahma*	See *Prajāpaṫi*. Also means, one who is the supervisor and coordinator of a *Yajṇa*.
ब्रह्मणः	*Brahmaṇa:*	See *Paramāṫma*.
ब्रह्मणाछन्सिः	*Brahmaṇāçansi:*	One of the three assistants of the *Brahma*.
ब्राह्मणम्	*Brāhmaṇam*	A section of the *Veḋa* dealing with rituals.
भारतम्	*Ḇāraṫam*	India and/or the Indian subcontinent comprising of Afghanistan, Bangladesh, Bhutan, India, Maldives, Myanmar, Nepal, Pakistan, Sri Lanka, and Tibet.
भाष्यं	*Ḇāṣya*	Commentary.
भूमि	*Ḇūmi*	Earth. Also, can be used to mean one of the five *Paṇcaḇūṫās*, namely the earth element. The same word is used for the *Ḋevaṫā* associated with Earth.
भ्राजः	*Ḇrāja:*	One of the seven *Sapṫasūryās*.

Devanāgarī	*Devanāgarī* **Transliterated**	**Word Meaning**
मन्त्रः	*Mantra:*	Within the greater *Vedic* tradition the word has a complex set of meanings and interpretations. In this book used to mean section of a *Prapāṭaka.*
मण्डलं	*Maṇdala*	Section of the heavens.
मत्स्यपुराणः	*Matsya Purāṇa:*	One of the *Purāṇas.*
मनसं	*Manasa*	Mind.
मन्त्रपुश्पम्	*Mantrapuṣpam*	22nd *Anuvāka* of *Tettirīya Āraṇyakam.*
मयूखाः	*Mayūkā:*	Rays.
मरुतः	*Maruta:*	One of the members of a class of *Devatās* associated with wind.
महत्	*Mahat*	See *Buddi.*
महाभारतम्	*Mahābāratam*	One of the two *Itihāsas.*
महामेरुः	*Mahāmeru:*	Considered to be the center of all the physical, metaphysical and spiritual universes.
महावेदि	*Mahāvedi*	The eastern section of the Vihara built for complex *Vedic* rituals such as *Somayāga.* It contain the *Uttaravedi*, *Sadasa*, and *Havirdāna.*

Devanāgarī	***Devanāgarī*** **Transliterated**	**Word Meaning**
मित्रः	*Mitra:*	See *Sūryadevata.*
मीमांसकाः	*Mīmā˙sakā:*	Scholars and followers of *Pūrvamīmāsā.*
मुहूर्तः	*Muhūrta:*	Unit of time, approximately equal to 1/30th of a day, which is measured as a time from one sunrise to the next.
मूर्तिः	*Mūrti:*	Symbolic physical embodiment of a *Devatā.*
यजमानः	*Yajamāna:*	The performer of the *Yajṉa.*
यज्ञः	*Yajṉa:*	*Vedic* fire ritual as per the strict prescribed rules. It is the foundation of the *Vedic* way of life and the core act of all *Vedic* rituals.
यज्ञपुरुषः	*Yajṉapuruṣa:*	A *Devatā* associated with *Yajṉa.*
यज्ञेनबन्धु	*Yajṉenabandu*	Entanglement through *Yajṉa.*
योगः	*Yoga:*	A group of physical, mental, and spiritual practices aimed at stilling the mind to know the self.

Devanāgarī	*Devanāgarī* **Transliterated**	**Word Meaning**
रथचक्रं	*Raṭacakra*	2 dimensional design on the ground for a *Vihāra* or its components, most importantly for the main altar of the *Yajña*.
रसाः	*Rasā:*	Bodily fluids.
राक्षसाः	*Rākṣasā:*	A class of beings who are human flesh eating demons.
राजा ऋतुपर्णः	*Rājā ṛtuparṇa:*	A character from *Mahābāratam*.
रुद्रः	*Rudra:*	*Devatā* associated with storms, wind, war, hunting and healing, amongst other things.
लोकाः	*Lokā:*	Worlds.
वज्रं	*Vajra*	Weapon of *Indra*.
वत्सः	*Vatsa:*	A *Ṛṣi*.
वराहवः	*Varāhava:*	One of the six *Vāyugaṇās*.
वरुणः	*Varuṇa:*	*Devatā* associated with waters, rivers, ocean, justice and moral law.
वरुणप्रघासः	*Varuṇapragāsa:*	One of the seasonal *Nemittikakarmans* performed during the rainy season.
वर्षा	*Varṣā*	Rain.

Devanāgarī	***Devanāgarī* Transliterated**	**Word Meaning**
वागम्भृणी	*Vāgamḇṛṇī*	*Devatā* associated with sound, speech and cosmic creative energy.
वातः	*Vāta:*	Air. Also see *Vāyu.*
वातराशनाः	*Vātarāṣanā:*	A family of *Ṛṣis.*
वायुः	*Vāyu:*	*Devatā* associated with wind and breath. Also can be used to mean one of the five *Pañcaḇūtās*, namely the air element.
वायुगणाः	*Vāyugaṇā:*	Energy field or some other type of field in the atmosphere.
वायुमण्डलम्	*Vāyumaṇḍalam*	Atmosphere.
वालखिल्याः	*Vālaḵilyā:*	A family of *Ṛṣis.*
वास्तुशास्त्रं	*Vāstuṣāstra* ˙	The ancient architectural science of *Ḇāratam.*
विध्युन्महसः	*Viḏyunmahasa:*	One of the six *Vāyugaṇās.*
विभासः	*Viḇāsa:*	One of the seven *Saptasūryās.*
विराटः	*Virāta:*	Anthropomorphological representation of the entire universe.
विश्वेदेवः	*Viṣvedeva:*	All *Devatās.*

Devanāgarī	*Devanāgarī* **Transliterated**	**Word Meaning**
विष्णुः	*Viṣṇu:*	*Devatā* associated with support and sustenance of heaven and earth among other things.
विहारः	*Vihāra:*	The temporary structures erected for the conduct of a *Vedic* ritual.
वेदः	*Veda:*	The sacred knowledge of the *Vedic* tradition. It is not authored by anyone, has always existed, will always exist, and is the main *Ṣabda Pramāṇā*.
वेदाङ्ग ·	*Vedāṅga·*	The six auxilliary disciplines, the study of which forms the foundation for study of the *Veda*.
वैखानसा:	*Vekānasā:*	A family of *Ṛṣis*.
वैशंपायनः	*Veṣa·pāyana:*	A *Ṛṣi*.
व्रतं	*Vrata·*	Various kinds of ritualistic observances in the greater *Vedic* tradition.
शब्दः	*Ṣabda:*	Knowledge through meaning testimony of past experts. One of the six *Pramāṇās*.
शमि	*Ṣami*	A type of a tree, Prosopis specigera.

Devanāgarī	***Devanāgarī* Transliterated**	**Word Meaning**
शम्बरउदकः	*Ṣambaraudaka:*	A type of water in the clouds that binds with sunlight to create rain.
शम्युः	*Ṣamyu:*	Son of *Bṛhaspati*.
शाखा	*Ṣākā*	A family that follows the *Vedic* tradition and specializes in the learning of certain sections of the *Veda*.
श्रुति	*Ṣruti*	See *Veda*.
श्रौतं	*Ṣrota˙*	See *Ṣrotasūtra*.
श्रौतसूत्रं	*Ṣrotasūtra˙*	One of the four sections of the *Kalpasūtrās*.
श्वापयः	*Ṣvāpaya*	One of the six *Vāyugaṇās*.
षोडशकर्मानि	*Ṣodaṣakarmāni*	See *Gṛhyakarman*.
संध्यावंदनं	*Sa˙ḍyāva˙dana˙*	A mandatory ritual that is performed at dawn and dusk.
संसकृतम्	*Sa˙sakṛtam*	Mother language of various Indic languages and the language of the *Vedic* tradition.
संहिता	*Sa˙hitā*	A section of the *Veda* consisting of a collection of *Sūktas*.
सत्यं	*Satya˙*	Truth.

Devanāgarī	*Devanāgarī* **Transliterated**	**Word Meaning**
सदसः	*Sadasa:*	The westernmost area of the *Vihāra* where additional supplemtary altars are built. It is also the place from where the *Udgātṛs* sing the *Vedic* Hymns.
सप्तकर्णः	*Saptakarṇa:*	A *Ṛṣi*.
सप्तसूर्याः	*Saptasūryā:*	The seven major components of the solar electromagnetic radiation spectrum.
सभ्यः	*Sabya:*	One of the five fires a *Śrota Gṛhasta* maintains.
सरस्वती	*Sarasvatī*	*Devatā* associated with knowledge, music, art, speech, wisdom, and learning and related things.
सलिलम्	*Salilam*	Source and cause of all the manifested
सवितरः	*Savitara:*	*Devatā* associated with the maintenance of *Ṛta*.
सायणाचार्यः	*Sāyaṇācārya:*	A *Mīmāsakā*, whose commentary of the *Veda* is the accepted commentary by both the practitioners and the acadamics.
सावित्रचयनं	*Sāvitracayana*	One of the *Cayanas*.

Devanāgarī	***Devanāgarī* Transliterated**	**Word Meaning**
सूक्तं	*Sūkta*	A collection of *Ṛcas*.
सूर्यदेवतः	*Sūryadevata:*	*Devatā* associated with the Sun.
सूर्यरश्मि	*Sūryaraṣmi*	Solar radiation.
सोमः	*Soma:*	*Devatā* associated with the *Soma* plant and all plant life in general. In the greater *Vedic* tradition *Soma* is associated with the moon.
सोमयागः	*Somayāga:*	The most complex *Nemittikakarman*, which must be performed at least once in one's lifetime.
स्तुति	*Stuti*	Praise.
स्मृति	*Smṛti*	*Purāṇās* and *Itihāsas*.
स्वतपसः	*Svatapasa:*	One of the six *Vāyugaṇās*.
स्वर्णरः	*Svarṇara:*	One of the seven *Saptasūryās*.
हविर्धानं	*Havirḍāna*	The area of the *Vihāra* east of the *Uttaravedi* and west of *Sadasa* where the two carts that bring the soma creepers are kept. It is the term used for the carts themselves as well.

Devanāgarī	*Devanāgarī* **Transliterated**	**Word Meaning**
हेतुः	*Hetu:*	Rational explaination based on established truths.
होताः	*Hotā:*	One who is responsible for the various invocations during a *Yajṇa*.

Glossary of Selected Scientific Terms

Abiogenesis: A hypothesis that life evolved from non-living matter.

Agnostic: A view that the existence of God(s) or such similar supernatural(s) is not certainly known.

Amino Acid: Fundamental structural units that make up proteins.

Anthropomorphic: The attributed human characteristics, emotions and/or behaviors to animals or inanimate things.

Apathetic: Not interested in or concerned about.

Atmosphere, Atmospheric Layers: Layers of gases that envelope a celestial body, and are held in place by that body's gravity.

Atomic Number: The number of protons found in the nucleus of an atom of any element.

Axial Tilt: The angle between the orbital plane and the axis of rotation of a celestial body.

Axis of Rotation: The diameter of a celestial body around which it rotates.

Baryogenesis: The hypothesis about the physical process that created matter at the time universe was formed.

Beryllium: A naturally occurring metallic element with an atomic number of 4.

Big Bang Theory: A hypothesis that explains the formation of the universe from an explosion followed by expansion of the explosive material.

Big Bounce: A hypothesis that explains the formation of the universe as a cyclical phenomenon of continuous expansion and contraction.

Biosphere: All living beings and their relationships, including their interaction with non-living matter.

Carbon: A naturally occurring non-metallic element with an atomic number of 6.

Catalyze: A catalyst is a substance that increases the rate of a chemical reaction without itself undergoing any permanent chemical change. This action of catalyst is the verb catalyze.

Cell, Cellular Life: Basic structural, functional and biological unit of all known organisms.

Cellular Metabolism: Chemical changes that take place in a cell through which energy and basic components are provided for essential processes.

Cellular Respiration: Process by which organisms combine oxygen with foodstuff molecules, diverting the chemical energy in these substances into life-sustaining activities and discarding, as waste products, carbon dioxide and water.

Chlorofluorocarbons: Gaseous compounds that contain carbon, chlorine, fluorine, and/or hydrogen, that are used as refrigerants, cleaning solvents, aerosol propellants or used in manufacture of plastic foams. They are a major cause of stratospheric ozone depletion.

Classical Newtonian Physics: Description of the physical events in the universe using the laws of motion and gravitation that were formulated by the physicist Sir Issac Newton.

Cloud Physics: The study of the physical processes underlying the formation, growth and precipitation of atmospheric clouds.

Cloud Micro Physics: The study of the various particles in atmospheric clouds.

Cognitive Science: The interdisciplinary study of the mind and its processes, namely cognition.

Collapse Models: The study of the emergence of the world as described by Classical Newtonian Physics as a particular case of the world as described by Quantum Physics.

Condensation: The change of the physical state of matter from the gas phase into the liquid phase.

Conductive Transfer: The transfer of internal energy by microscopic collisions of particles and movement of electrons within a body.

Consciousness: A yet not scientifically understood aspect of humans. It is loosely defined as human awareness of both internal and external world.

Core: The central part of celestial body.

Cosmic Radiation: High energy photons and other charged particles traveling at the speed of light from the sun and the stars from various galaxies.

Cosmic Ray Fission: Nuclear fission reactions ensuing from cosmic radiation.

Cosmogony: The study of the origin of the universe, especially the solar system.

Cosmology: The study of origin, structure, space-time relationships and metaphysics of the universe.

Crust: The outermost solid layer of rocky planets, natural satellites and similar celestial bodies.

Cryptography: Study of techniques for secure communication.

Cyanobacteria: Considered to be the earliest known form of life on the earth. They are a division of microorganisms that are related to the bacteria but are capable of photosynthesis.

Dark Energy: Hypothetical, and as of now undetectable energy in the universe. Its existence is postulated by indirect inference. It is supposed to account for 68% of all energy in the universe.

Dark Matter: Hypothetical, and as of now undetectable matter in the universe. Its existence is postulated by indirect inference. It is supposed to account for 85% of all matter in the universe.

Detection Efficiency: One of the loopholes in Quantum Entanglement tests, where the chosen sample of entangled pairs in the experiment may not be an accurate representation of the population.

Deuterium: Naturally occurring stable isotope of Hydrogen. Has a neutron in addition to a proton in the nucleus, whereas Hydrogen only has a Proton in its nucleus.

Deuterium Oxide: Water formed with Deuterium instead of Hydrogen.

Diurnal: During the day.

Diurnal Temperature Variation: The variation in air temperature during the day.

Dogmatic: Based on belief with/without any proof of that belief.

East-West Winds: The permanent east-to-west prevailing winds that flow in the Earth's equatorial region.

Ekpyrotic: A cosmological model of the universe that postulates that the universe undergoes a continuous cycle of expansion and contraction.

Electromagnetic Force: The physical interaction that occurs between electrically charged particles.

Electromagnetic Radiation: A wave of synchronized oscillations of electric and magnetic fields. Alternatively they can be viewed as a stream of photons, which are uncharged but energized elementary particles with zero rest mass. Examples include visible light, microwaves, radio waves, infrared rays, x-rays, ultraviolet rays, gamma rays etcetera.

Electron Cloud: The system of electrons surrounding the nucleus of an atom.

Electronic Transition Frequency: The frequency at which electrons jump from one energy level to another energy level in an atom.

Electrostatic Forces: Electrostatic forces are attractive or repulsive forces between particles that are caused by their electric charges.

Elementary Particle: A subatomic particle is that is not composed of other particle.

Empirical: That which is based on, related to, or verifiable by observation or experience instead of theory and/or logic.

Energy Field: A field of an electromagnetic charge that exerts force on other charges.

Epistemology: The study of sources of knowledge, nature of knowledge and scope of knowledge, and related issues.

Euclidean Space: The fundamental space of classical geometry.

Evaporate, Evaporation: Vaporization is the process by which a material in its liquid form changes in to this gaseous form. This phenomena occurring on the surface of the liquid body is called evaporation.

Exegesis: A critical interpretation of a text, particularly a religious text.

Falsifiability: A statement is falsifiable if there is a possibility that it can be proven false.

Field: A physical quantity, represented by a number or another complex metric such as a Tensor, that has a value for each point in space and time.

Frequency: Number of occurrences of an event in a unit of time.

Gaussian Curves: The characteristic symmetrical bell curves that are generated from a mathematical function called the Gaussian Function.

General Theory of Relativity: A theory, which describes gravity as a geometric property of space and time.

Gluons: Massless elementary particles that stick other particles called Quarks together to form subatomic particles such as protons and neutrons.

Gravitational Collapse: Contraction of the material constituting a celestial body inwards to its center of gravity.

Gravitational Force: Force as per Classical Newtonian Physics that causes Gravity.

Graviton: Hypothetical elementary particle which is believed to mediate the force of gravitational interaction.

Gravity: A natural phenomenon by which all objects with mass and energy in the universe are attracted to one another.

Gravity Waves: Waves generated in a fluid medium or at the interface between two media when the force of gravity or buoyancy tries to restore equilibrium.

Greenhouse Effect: The process by which radiation from a planet's atmosphere warms the planet's surface to a temperature above what it would be without this atmosphere

Hadron: A composite particle made up of two or more elementary particles called quarks.

Halons: Gaseous compounds of carbon formed by combination with bromine and other halogens.

Heavier Elements: Elements with atomic number greater than 26, which is the atomic number of Iron (Ferrum FE).

Helium: Colorless, odorless, tasteless, non-toxic, inert, monatomic gas element with an atomic number of 2.

Hertz: The standard international unit of frequency, equal to one cycle per second.

Human Anatomy: The branch of biology that deals with the form of human body, and with relationships between the various structures within it.

Human Physiology: The branch of biology that deals with the study of the mechanical, physical, and biochemical functions of the human body.

Hydrogen: Colorless, odorless, non-toxic, and highly combustible diatomic gas element with an atomic number of 1.

Hydrosphere: The combined water mass on the surface, below the surface, and above the surface of Earth or any other celestial body.

Hydrothermal Vent Theory: Theory that posits that life may have begun at submarine hydrothermal vents.

Hypothesis: A proposed explanation of a phenomena.

Infiltration: Water on the ground surface enters the soil.

Inorganic Matter: Non-living matter that does not have any molecules having carbon and Hydrogen.

Inter Specie Evolution: A hypothesis of how higher species of life evolved from lower species.

Interstellar Gas and Dust: The subset of interstellar medium, which includes cosmic radiation, dust, atomic, molecular and ionic gas that fills the space between various star systems in the universe.

Iron: A naturally occurring metallic element with an atomic number of 26.

Isotopes: Two different types of atoms of the same element having the same atomic number but different atomic mass due to differing number of neutrons in their nuclei.

Lambda CDM Parametrization: A hypothetical parametrized big bang cosmological model according to which the universe mainly contains , dark energy denoted by the Greek letter Lambda (Λ), dark matter (postulated as cold dark matter and abbreviated as CDM), and ordinary matter.

Latitude: A variable that specifies the position of any point on the earth's surface in the North-South direction. It is expressed as an angular measure between the lines that connect the North Pole and South Pole with the radial line connecting the point in discussion with the center of the Earth.

Laws of Motion: The three Classical Newtonian Mechanics laws that describe the relationship between the motion of an object and the forces acting on it.

Lighter Elements: Elements with atomic number lesser than or equal to 26, which is the atomic number of Iron (Ferrum FE).

Lithium: A naturally occurring metallic element with an atomic number of 3.

Lithosphere: The rigid outer part of the earth or other celestial bodies, consisting of the crust and upper mantle.

Local Hidden Variables: Certain yet to be discovered variables thought to cause quantum entanglement.

Locality: One of the loopholes in Quantum Entanglement tests in which the particle separation sites are not far enough.

Longitude: A variable that specifies the position of any point on the earth's surface in the East-West direction. It is expressed as an angular measure between the radial line from a point on the prime meridian to the center of the earth with the radial line connecting the point in discussion with the center of the Earth. The chosen point on the prime meridian to measure the longitude is in the same latitude as the point in discussion. The prime meridian is a curved line on the earth's surface connecting the north pole and south pole and, which passes near the Royal Observatory, Greenwich, England, is defined as 0° longitude by convention.

Loop Quantum Gravity: Theory of quantum gravity, which aims to merge quantum mechanics and general relativity.

Luminosity: The radiant power emitted by a light-emitting object over time or the total amount of electromagnetic energy emitted per unit of time by a star, galaxy, or other celestial object.

Macrocosm: The universe as a whole. See Microcosm.

Magnetosphere: The region surrounding the earth or another celestial body in which its magnetic field is the predominant effective magnetic field

Mantle: Usually the largest layer of a planetary or similar celestial bodies. It is the layer between the outer crust and inner core of the body.

Metaphysics: A philosophical attempt to describe reality beyond the scope of existing scientific knowledge.

Meteorite: A solid piece of debris from outer space that survives its passage through the atmosphere to reach the surface of Earth or similar celestial body.

Microcosm: Human related spaces including human bodies, which are believed to be miniature reflections of the Macrocosm with a structural and/or functional similarity between them.

Microwaves: Electromagnetic radiation with wavelengths ranging from about one meter to one millimeter and corresponding to frequencies between 300 MHz and 300 GHz respectively.

Molecular Mass: Mass of a molecule.

Natural Sciences: Branch of science concerned with the study of natural phenomena.

Nebular Hypothesis: A hypothesis that explains the formation and evolution of the solar system according to which the solar system originated from the gas and dust floating around the Sun.

Noctilucent Clouds: Tenuous cloud type phenomena made up of micro ice crystals in the upper atmosphere of Earth visible only during twilight hours.

Nuclei: The positively charged central portion of an atom that comprises nearly all of the atomic mass and that consists of protons and neutrons.

Nucleogenesis: A process by which new atomic nuclei are created from existing atomic nuclei, protons and neutrons.

Organic: Related to or derived from living matter primarily made of molecules having carbon and hydrogen.

Organism: Organic living system that functions as an independent entity.

Orthodoxical, Orthodoxy: Importance of codified beliefs in form of creeds as opposed to correct conduct and action.

Orthopraxical, Orthopraxy: Importance of correct conduct and action as opposed to codified beliefs in form of creeds.

Oxygen: A naturally occurring a colorless, odorless, and highly reactive diatomic gas with an atomic number of 8.

Ozone: A pale blue triatomic oxygen gas with a distinctively pungent smell formed naturally in greatest concentrations in the ozone layer of the stratosphere from the interactions of diatomic oxygen and the ultraviolet radiation.

Ozone Disassociation: The breaking of the ozone molecule into diatomic oxygen molecule and monoatomic oxygen.

Ozone-Oxygen Cycle: The continuous conversion of diatomic oxygen into ozone and vice versa.

Panspermia Theory: Hypothesis that life exists throughout the Universe.

Parametrization: A mathematical process consisting of expressing the state of a system, process or model as a function of some independent quantities called parameters.

Particle: Smallest countable physical object as per physicists.

Particle Physics: A branch of physics that studies the nature of the irreducibly smallest detectable particles that constitute matter and radiation.

Photo Ionization: Th process by which ions are created from the interaction of photons with atoms or molecules.

Photoelectric Effect: The process by which electrons are emitted when electromagnetic radiation hits a material.

Photon: A particle representing a quantum of light or other electromagnetic radiation.

Photosynthesis: The process by which green plants and some other organisms use sunlight to synthesize food and oxygen from carbon dioxide and water.

Planetary Waves: Naturally occurring waves in the oceans and the atmosphere due to the rotation of the earth.

Planetesimals: Solid objects posited to exist in protoplanetary disks.

Postulate: See Hypothesis.

Potentiality: Latent qualities or abilities that have not manifested yet.

Precipitation: Condensation of water vapor in the atmosphere and its fall to the earth's surface due to gravity. It includes drizzle, rain, snow, sleet, ice pellets, graupel and hail.

Primordial: Existing at or beginning of time. In creative cycle theories it marks the onset of each new cycle of creation.

Protoplanetary Disk: A rotating torus, pan cake or ring shaped disc of dense gas and dust surrounding a new formed star.

Quantum Entanglement: Quantum entanglement is a physical phenomenon in which two or more particles become related such that the quantum states of the entangled particles cannot be described independently of the other particles in the entangled group, even when these particles are separated at great distances.

Quantum Physics: A branch of physics that studies the physical properties of nature at the scale of atoms and subatomic particles.

Quantum State: Quantum state is a mathematical entity that provides a probability distribution for the outcomes of each possible measurement on a system.

Quantum: A quantum is the minimum amount of any physical entity involved in an interaction.

Quantum Bridge: A bridge between contracting and expanding cosmological branches in a cyclical cosmological model.

Quantum Computing: The use of collective properties of quantum states to do computations

Quantum Cryptography: The science of utilizing quantum mechanical properties to perform cryptographic tasks.

Quantum Decoherence: Loss of quantum entanglement.

Quantum Field Collapse: See Quantum Decoherence.

Quantum Fields: A structure used to describe the behavior of particles using probability distributions.

Quantum Geometry: The set of mathematical concepts generalizing the concepts of geometry whose understanding is necessary to describe the physical phenomena at distance scales comparable to the Planck length (Approximately $1.616255\times10-35$ meters)

Quantum Gravity: A field of theoretical physics that seeks to describe gravity according to the principles of quantum physics.

Quantum Teleportation: Technique for transferring quantum information from a sender at one location to a receiver some distance away.

Quantum Tunneling: The quantum mechanical phenomenon where a wavefunction can propagate through a potential barrier.

Quarks: A quark is a type of elementary particle and a fundamental constituent of matter.

Radiometric Dating: A technique of dating materials by determining the relative proportions of certain trace radioactive substances present in them.

Repeatability: The degree of repeatability for a given experiment is determined by how close the results of an experiment are in repeated trials of that experiment.

Revolution: The movement of a celestial body or a celestial system around another celestial body or center of a celestial system.

Runoff: The portion of the precipitation that does not infiltrate into the soil but flows on the surface.

Singularity: A hypothesized entity of infinite density containing all mass, energy, and space-time. This entity is postulated as the seed from which the universe is born.

Solar Electromagnetic Radiation: Electromagnetic radiation emanating form the Sun.

Solar Nebula: See Nebular Hypothesis.

Solar System: The sun and all the celestial bodies that directly or indirectly revolve around it.

Spacetime Fabric: An imaginary fabric in which everything in the universe is thought to be embedded as per Einstein's Theory of Relativity.

Standard Model: The theoretical description of three of the four known fundamental forces (the electromagnetic, weak interaction, and strong interaction, while omitting gravity) in the universe, as well as classifying all known elementary particles.

Strong Nuclear Force, Strong Interaction: The force inside the nucleus of an atom which binds the neutrons and protons together creating the nucleus and that which binds the quarks to form the protons, nucleons and other hadrons themselves.

Sublimate: The direct change from solid to gaseous state.

Supernovae: A powerful explosion of a star generating enormous luminosity.

Synodic Rotation: A rotation time of a celestial body around its own axis with reference to the celestial body or system around which the rotating celestial body is revolving.

Theistic, Theism: Belief in the existence of Gods or God.

Theory of Evolution: The theory that explains the evolution of species.

Thermonuclear Fusion: The process in which two or more atomic nuclei combine to form one or more different atomic nuclei and subatomic particles and the difference in mass before and after the combination is released as heat energy.

Transpiration: Evaporation of water from the plants via their aerial parts including stems, leaves and flowers.

Tritium: A radioactive isotope of Hydrogen with one proton and two neutrons in its nucleus.

Velocity: the rate of change of position along a straight line with respect to time : the derivative of position with respect to time

Water Cycle: The continuous movement of water above and below the earth's surface and through its lower atmosphere via the transpiration, evaporation, condensation, precipitation, runoff, and infiltration processes.

Weak Nuclear Force, Weak Interaction: The mechanism of interaction between subatomic particles that is responsible for the radioactive decay of atoms.

Text of *Aruṇapraśna:*

Reference	Text *Devanāgarī*
1.1.1	ॐ भद्रं कर्णेभिः श्रृणुयाम देवाः । भद्रं पश्येमाक्षभिर्यजत्राः । स्थिरैरङ्गैस्तुष्टुवाꣳ सस्तनूभिः । व्यशेम देवहितं यदायुः । स्वस्ति न इन्द्रो वृध्दश्रवाः । स्वस्ति नः पूषा विश्ववेदाः । स्वस्तिनस्ताक्ष्र्यो अरिष्टनेमिः । स्वस्ति नो बृहस्पतिर्दधातु ॥
1.1.2	ॐ भद्रं कर्णेभिः श्रृणुयाम देवाः । भद्रं पश्येमाक्षभिर्यजत्राः । स्थिरैरङ्गैस्तुष्टुवाꣳ सस्तनूभिः । व्यशेम देवहितं यदायुः ॥
1.1.3	स्वस्ति न इन्द्रो वृध्दश्रवाः । स्वस्ति नः पूषा विश्ववेदाः । स्वस्तिनस्ताक्ष्र्यो अरिष्टनेमिः । स्वस्ति नो बृहस्पतिर्दधातु ॥
1.1.4	आपमापामपः सर्वाः । अस्मादस्मादितोऽमुतः । अग्निर्वायुश्च सूर्यष्च । सह संचस्करर्द्धिया ॥
1.1.5	वाय्वश्वा रश्मिपतयः । मरीच्यात्मानो अद्रुहः । देवीर्भुवनसूवरीः । पुत्रवत्वाय मे सुत ॥
1.1.6	महानाम्नीर्महामानाः । महसो महसस्स्वः । देवीः पर्जन्यसूवरीः । पुत्रवत्वाय मे सुत ॥
1.1.7	अपाश्न्युष्णिमपा रक्षः । अपाश्न्युष्णिमपारघम् । अपाघ्रामप चावर्तिम् । अपदेवीरितो हित ॥
1.1.8	वज्रं देवीरजीताꣳश्च । भुवनं देवसूवरीः । आदित्यानदितिं देवीम् । योनिनोर्ध्वमुदीषत ॥
1.1.9	शिवानश्शन्तमा भवन्तु । दिव्या आप ओषधयः ॥
1.1.10	सुमृडीका सरस्वती । मा ते व्योम संदृशि ॥
1.2.1	स्मृतिः प्रत्यक्षमैतिह्यम् । अनुमानश्चतुष्टयम् । एतैरादित्यमण्डलम् । सर्वैरेव विद्यास्यते ॥
1.2.2	सूर्यो मरीचिमादत्ते । सर्वस्माद्भुवनादधि । तस्याः पाकविशेषेण । स्मृतं कालविशेषणम् ॥

Reference	Text *Devanāgarī*
1.2.3	नदीव प्रभवात्काचित् । अक्षय्यात्स्यन्दते यथा ॥ तान्नध्योऽभिसमायन्ति । सोरुस्सती न निवर्तते ॥
1.2.4	एवन्नानासमुत्थानाः । कालास्संवत्सरग्ग् श्रिताः । अणुशश्च महशश्च । सर्वे समवयन्त्रितम् ॥
1.2.5	सतैस्सर्वैस्समाविष्टः । ऊरुस्संन्न निवर्तते । अधिसंवत्सरं विध्यात् । तदेव लक्षणे ॥
1.2.6	अणुभिश्च महद्भिश्च । समारूढः प्रदृश्यते । संवत्सरः प्रत्यक्षेण । नाधिसत्त्वः प्रदृश्यते ॥
1.2.7	पटरो विक्लिधः पिङ्गः । एतद्वरुणलक्षणम् । यत्रैतदपुदृश्यते । सहस्रं तत्र नीयते ॥
1.2.8	एकꣳहि शिरो नाना मुखे । कृत्स्नं तदृतुलक्षणम् । उभयतस्सप्तेन्द्रियाणि । जल्पितं त्वेव दिह्यते ॥
1.2.9	शुक्लकृष्णे संवत्सरस्य । दक्षिणवामयोः पार्श्वयोः । तस्यैषा भवति ॥
1.2.10	शुक्रं ते अन्यद्यजतं ते अन्यत् । विषुरूपे अहनी द्यौरिवासि । विश्वा हि माया अवसि स्वधावः । भद्रा ते पूषन्निह रातिरस्त्विति ॥
1.2.11	नात्र भुवनम् । न पूषा । न पशवः । नादित्यस्संवत्सर एव प्रत्यक्षेण प्रियतमं विद्यात् । एतद्वै संवत्सरस्य प्रियतमꣳ रूपम् । योऽस्य महानर्थ उत्पत्स्यमानो भवति । इदं पुण्यं कुरुष्वेति । तमाहरणं दद्यात् ॥
1.3.1	साकंजानाꣳ सप्तथमाहुरेकजम् । षडिद्यमा ऋषयो देवजा इति । तेषामिष्टानि विहितानि धामशः । स्थात्रे रेजन्ते विकृतानि रूपशः ॥
1.3.2	को नु मर्या अमिथितः । सखा सखायमब्रवीत् । जहाको अस्मदीषते ॥

Reference	Text *Devanāgarī*
1.3.3	यस्तित्याजं सखिविदꣳ सखांयाम् । न तस्यं वाच्यपिं भागों अंस्ति । यदींꣳ शृणोत्यलकꣳ शृणोति । न हि प्रवेद सुकृतस्य पन्थामितिं ॥
1.3.4	ऋतुर्ऋतुना नुद्यमांनः । विनंनादाभिधांवः । षष्टिश्च त्रिꣳशंका वल्गाः । शुक्लकृष्णौ च षाष्टिंकौ ॥
1.3.5	सारागवस्त्रैर्जरदक्षः । वसन्तो वसुंभिस्सह । संवत्सरस्यं सवितुः । प्रैषकृत्प्रंथमः स्मृतः ॥
1.3.6	अमूनादयंतेत्यन्यान् । अमूग्श्चं परिरक्षंतः । एता वाचः प्रंयुज्यंन्ते । यत्रैतंदपुदृश्यंते ॥
1.3.7	एतदेव विंजानीयात् । प्रमाणं कालपर्ययें । विशेषणं तुं वक्ष्यामः । ऋतूनां तन्निबोधंत ॥
1.3.8	शुक्लवासां रुद्रगणः । ग्रीष्मेणांऽऽवर्तते संह । निजहंन्पृथिंवीꣳ सर्वाम् । ज्योतिषांऽप्रतिख्येनं सः ॥
1.3.9	विश्वरूपाणिं वासाꣳसि । आदित्यानां निबोधंत । संवत्सरीणं कर्मफलम् । वर्षाभिर्ददताꣳ संह ॥
1.3.10	अदःखों दःखचंक्षुरिव । तद्माऽऽपीत इव दृश्यंते । शीतेनंव्यथंय न्निव । रुरू दंक्षइ वदृश्यंते ॥
1.3.11	ह्लादयतें ज्वलंतश्चैव । शाम्यतंश्चास्य चक्षुंषी । या वै प्रजा भ्रंग्श्यन्ते । संवत्सरत्ता भ्रंग्श्यन्ते ॥
1.3.12	याः प्रतिंतिष्ठन्ति । संवत्सरे ताः प्रतिंतिष्ठन्ति । वर्षाभ्यं इत्यर्थः ॥
1.4.1	अक्षिंदःखोत्थिंतस्यैव । विप्रसंन्ने कनीनिंके । आङ्क्ते चाद्गंणं नास्ति । ऋभूणां तन्निबोधंत ॥
1.4.2	कनकाभानिं वासाꣳसि । अहतांनि निबोधंत । अन्नमश्नीतं मृज्मीतं । अहं वों जीवनप्रंदः ॥
1.4.3	एता वाचः प्रंयुज्यन्ते । शरद्यंत्रोपदृश्यंते । अभिधून्वन्तोऽभिघ्नंन्त इव । वातंवन्तो मरुद्गंणाः ॥

Reference	Text *Devanāgarī*
1.4.4	अमुतो जेतुमिषुमुखमिव । सन्नध्दास्सह ददृशे ह । अपध्वस्तैर्वस्तिवर्णैरिव । विशिखासः कपर्दिनः ॥
1.4.5	अक्रुध्दस्य योत्स्यमानस्य । क्रुध्दस्येव लोहिनी । हेमतश्चक्षुषी विद्यात् । अक्षणयोः क्षिपणोरिव ॥
1.4.6	दर्भिक्षं देवलोकेषु । मनूनामुदकं गृहे । एता वाचः प्रवदन्तीः । वैद्युतो यान्ति शैशिरीः ॥
1.4.7	ता अग्निः पवमाना अन्वैक्षत । इह जीविकामपरिपश्यन् । तस्यैषा भवति ॥
1.4.8	इहेह वस्स्वतपसः । मरुतस्सूर्यत्वचः । शर्म सप्रथा आवृणे ॥
1.5.1	अतिताम्राणि वासाꣳसि । अष्टिवज्रिशतघ्नि च । विश्वे देवा विप्रहरन्ति । अग्निजिह्वा असश्चत ॥
1.5.2	नैव देवो न मर्त्यः । न राजा वरुणो विभुः । नाग्निर्नेन्द्रो न पवमानः । मातृक्कचन विद्यते ॥
1.5.3	दिव्यस्यैका धनुरार्तिः । पृथिव्यामपरा श्रिता ॥ तस्येन्द्रो वम्रिरूपेण । धनुर्ज्यामछिनथ्स्वयम् ॥
1.5.4	तदिन्द्रधनुरित्यज्यम् । अभ्रवर्णेषु चक्षते । एतदेव शंयोर्बार्हस्पत्यस्य । एतद्रुद्रस्य धनुः ॥
1.5.5	रुद्रस्य त्वेव धनुरार्तिः । शिर उत्पिपेष । स प्रवर्ग्योऽभवत् । तस्माद्यस्सप्रवर्ग्येण यज्ञेन यजते । रुद्रस्य स शिरः प्रतिदधाति । नैनꣳरुद्र आरुको भवति । य एवं वेद ॥
1.6.1	अत्यूर्ध्वाक्षोऽतिरश्चात् । शिशिरः प्रदृश्यते । नैव रूपं न वासाꣳसि । न चक्षुः प्रतिदृश्यते ॥
1.6.2	अन्योन्यं तु न हिग्ग्स्नातः । सतस्तद्देवलक्षणम् । लोहितोऽक्षिण शारशीर्ष्णिः । सूर्यस्योदयनं प्रति ॥

Reference	Text *Devanāgarī*
1.6.3	त्वं करोषिन्न्यञ्जलिकाम् । त्वं करोषि निजानुकाम् । निजानुका मेन्न्यञ्जलिका । अमी वाचमुपासतामिति ॥
1.6.4	तस्मै सर्व ऋतवो नमन्ते । मर्यादाकरत्वात्प्रपुरोधाम् । ब्राह्मणं आप्नोति । य एवं वेद । स खल संवत्सर एतैस्सेनानीभिस्सह । इन्द्राय सर्वान्कामानभिवहति । स द्रप्सः । तस्यैषा भवति ॥
1.6.5	अवद्रप्सो अꣳशुमतीमतिष्ठत् । इयानः कृष्णो दशभिः सहस्रैः । आवर्तमिन्द्रः शच्या धमन्तम् । उपस्नुहि तं नृमणामथद्रामिति ॥
1.6.6	एतयैवेन्द्रः सलावृक्या सह । असुरान्परिवृश्चति । पृथिव्यꣳशुमती । तामन्ववस्थितः संवत्सरो दिवं च । नैवंविदुषाऽऽचार्यान्तेवासिनौ । अन्योन्यस्मै द्रुह्याताम् । यो द्रुह्यति । भ्रश्यते स्वर्गाल्लोकात् । इत्यृतुमण्डलानि । सूर्यमण्डलान्याख्यायिकाः । अत ऊर्ध्वꣳसनिर्वचनाः ॥
1.7.1	आरोगो भ्राजः पतरः पतङ्गः । स्वर्णरो ज्योतिषीमान् विभासः । ते अस्मै सर्वे दिवमातपन्ति । ऊर्जं दुहाना अनपस्फुरन्त इति ॥
1.7.2	कश्यपोऽष्टमः । स महामेरुं न जहाति । तस्यैषा भवति ॥
1.7.3	यत्ते शिल्पं कश्यप रोचनावत् । इन्द्रियावत्पुष्कलं चित्रभानु । यस्मिन्सूर्या अर्पितास्सप्त साकम् ॥ तस्मिन्राजानमधिविश्रयेममिति ॥
1.7.4	ते अस्मै सर्वे कश्यपाज्ज्योतिर्लभन्ते । तान्सोमः कश्यपादधिनिर्ध्दमति । भ्रस्ताकर्मकृदिवैवम् ॥
1.7.5	प्राणो जीवानीन्द्रियंजीवानि । सप्त शीर्षण्याः प्राणाः । सूर्या इत्याचार्याः ॥
1.7.6	अपश्यमहमेतन्सप्त सूर्यानिति । पञ्चकर्णो वात्स्यायनः । सप्तकर्णश्च प्लाक्षिः । आनुश्रविक एव नौ कश्यप इति । उभौ वेदयिते । न हि शेकुमिव महामेरुं गन्तुम् ॥

Reference	**Text *Devanāgarī***
1.7.7	अपश्यमहमेतत्सूर्यमण्डलं परिवर्तमानम् । गार्ग्यः प्राणत्रातः । गच्छन्त महामेरुम् । एकं चाजहतम् ॥
1.7.8	भ्राजपतरपतङ्गा निहने ॥ तिष्ठन्नातपन्ति । तस्मादिह तप्त्रितपाः ॥
1.7.9	तेषामेषा भवति ॥
1.7.10	सप्त सूर्या दिवमनुप्रविष्टाः । तानन्वेति पथिभिर्दक्षिणावान् । ते अस्मै सर्वे घृतमातपन्ति । ऊर्जं दुहाना अनपस्फुरन्त इति ॥
1.7.11	सप्तर्त्विजसूर्या इत्याचार्याः ॥
1.7.12	तेषामेषा भवति ॥
1.7.13	सप्त दिशो नानासूर्याः । सप्त होतार ऋत्विजः । देवा आदित्या ये सप्त । तेभिः सोमाभीरक्षण इति ॥
1.7.14	तदप्याम्नायः । दिग्भ्राज ऋतून् करोति ॥
1.7.15	एतयैवावृताऽऽसहस्रसूर्यताया इति वैशंपायनः ॥
1.7.16	तस्यैषा भवति ॥
1.7.17	यद्याव इन्द्र ते शतꣳशतं भूमीः । उत स्युः । नत्वा वज्रिन्सहस्रꣳसूर्याः । अनुनजातमष्ट रोदसी इति ॥
1.7.18	नानालिङ्गत्वादृतूनां नानासूर्यत्वम् ॥
1.7.19	अष्टौ तु व्यवसिता इति ॥
1.7.20	सूर्यमण्डलान्यष्टात ऊर्ध्वम् ॥
1.7.21	तेषामेषा भवति ॥

Reference	Text *Devanāgarī*
1.7.22	चित्रं देवानामुदगादनीकम् । चक्षुर्मित्रस्य वरुणस्याग्नेः । आप्रा द्यावापृथिवी अन्तरिक्षम् । सूर्य आत्मा जगतस्तस्थुषश्चेति ॥
1.8.1	क्वेदमभ्रन्निविशते । क्वायꣳ संवत्सरो मिथः । क्वाहः क्वेयंन्देवं रात्री । क्व मासा ऋतवः श्रिताः ॥
1.8.2	अर्धमासा मुहूर्ताः । निमेषास्त्रुटिभिस्सह । क्वेमा आपो निविशन्ते । यदीतो यान्ति संप्रति ॥
1.8.3	काला अप्सु निविशन्ते । आपस्सूर्ये समाहिताः । अभ्राण्यपः प्रपद्यन्ते । विद्युत्सूर्ये समाहिता ॥
1.8.4	अनवर्णे इमे भूमी । इयं चासौ च रोदसि ॥
1.8.5	किग्गिस्वदत्रान्तरा भूतम् । येनेमे विधृते उभे । विष्णुणां विधृते भूमी । इति वत्सस्य वेदना ॥
1.8.6	इरावती धेनुमती हि भूतम् । सूयवसिनी मनुषे दशस्यै । व्यष्टभ्नाद्रोदसी विष्णवेते । दाधर्थ पृथिवीमभितो मयूखैः ॥
1.8.7	किन्तद्विष्णोर्बलमाहुः । का दीप्तिः किं परायणम् । एको यध्दारयद्देवः । रेजती रोदसी उभे ॥
1.8.8	वाताद्विष्णोर्बलमाहुः । अक्षराद्दीप्तिरुच्यते । त्रिपदाध्दारयद्देवः । यद्विष्णोरेकमुत्तमम् ॥
1.8.9	अग्नयो वायवश्चैव । एतदस्य परायणम् ॥
1.8.10	पृच्छामि त्वा परं मृत्युम् । अवमं मध्यमश्चतुम् । लोकश्च पुण्यपापानाम् । एतन्पृच्छामि संप्रति ॥
1.8.11	अमुमाहुः परं मृत्युम् । पवमानं तु मध्यमम् । अग्निरेवावमो मृत्युः । चन्द्रमाश्चतुरुच्यते ॥

Reference	Text *Devanāgarī*
1.8.12	अनाभोगाः परं मृत्युम् । पापास्संयन्ति सर्वदा । आभोगास्त्वेव संयन्ति । यत्र पुण्यकृतो जनाः ॥
1.8.13	ततो मध्यममायन्ति । चतुमंग्निं च संप्रति ॥
1.8.14	पृच्छामि त्वां पापकृतः । यत्र यांतयते यमः । त्वं नस्तद्ब्रह्मन् प्रब्रूहि । यदि वेत्थासतो गृहान् ॥
1.8.15	कश्यपादुदितास्सूर्याः । पापान्निघ्नन्ति सर्वदा । रोदस्योरन्तदेशेषु । तत्र न्यस्यन्ते वासवैः ॥
1.8.16	तेऽशरीराः प्रपद्यन्ते । यथाऽपुण्यस्य कर्मणः । अपाण्यपादकेशासः । तत्र तेऽयोनिजा जनाः ॥
1.8.17	मृत्वा पुनर्मृत्युमापद्यन्ते । अद्यमानास्स्वकर्मभिः । आशातिकाः क्रिमय इव । ततः पूयन्ते वासवैः ॥
1.8.18	अपैतं मृत्युं जयति । य एवं वेद । स खल्वेवंविद्ब्राह्मणः । दीर्घश्रुत्तमो भवति । कश्यपस्यातिथिस्सिद्धिगमनस्सिद्धागमनः ॥
1.8.19	तस्यैषा भवति ॥
1.8.20	आ यस्मिन्थ्सस वासवाः । रोहन्ति पूर्व्या रुहः । ऋषिर्ह दीर्घश्रुत्तमः । इन्द्रस्य घर्मो अतिथिरिति ॥
1.8.21	कश्यपः पश्यको भवति । यत्सर्वं परिपश्यतीति सौक्ष्म्यात् ॥
1.8.22	अथाग्नेरष्टपुरुषस्य । तस्यैषा भवति ॥
1.8.23	अग्ने नय सुपथा राये अस्मान् । विश्वानि देव वयुनानि विद्वान् । युयोध्यस्मञ्जुहुराणमेनः । भूयिष्ठान्ते नम उक्तिं विधेमेति ॥

Reference	**Text *Devanāgarī***
1.9.1	अग्निश्च जातवेदाश्च । सहोजा अजिराप्रभुः ।, वैश्वानरो नर्यापाश्च । पङ्क्तिराधाश्च सप्तमः । विसर्पेवाऽष्टमोऽग्नीनाम् । एतेऽष्टौ वसवः क्षिता इति ॥
1.9.2	यथर्त्वेवाग्नेरर्चिर्वर्णविशेषाः । नीलार्चिश्च पीतकार्चिश्चेति ॥
1.9.3	अथ वायोरेकादशपुरुषस्यैकादशस्त्रीकस्य ॥
1.9.4	प्रभ्राजमाना व्यवदाताः । याश्च वासुकिवैद्युताः । रजताः परुषाः श्यामाः । कपिला अतिलोहिताः । ऊर्ध्वा अवपतन्ताश्च । वैद्युत इत्येकादश ॥
1.9.5	नैनं वैद्युतो हिनस्ति । य एवं वेद ॥
1.9.6	स होवाच व्यासः पाराशर्यः । य एवं वेद । स होवाच व्यासः पाराशर्यः । विद्युद्वधमेवाहं मृत्युमैच्छमिति ॥
1.9.7	न त्वकामꣳहन्ति । य एवं वेद ॥
1.9.8	अथ गन्धर्वगणाः ॥
1.9.9	स्वानभ्राट । अङ्घारिर्बम्भारिः । हस्तस्सुहस्तः । कृशानुर्विश्वावसुः । मूर्धन्वास्थ्सूर्यवर्चाः । कृतिरित्येकादश गन्धर्वगणाः ॥
1.9.10	देवाश्च महादेवाः । रश्मयश्च देवा गरगिरः ॥
1.9.11	नैनं गरो हिनस्ति । य एवं वेद ॥
1.9.12	गौरीमिमाय सलिलानि तक्षती । एकपदी द्विपदी सा चतुष्पदी । अष्टपदी नवपदी बभूवुषी । सहस्राक्षरा परमे व्योमन्निति ॥
1.9.13	वाचो विशेषणम् ॥
1.9.14	अथ निगदव्याख्याताः ।ताननुक्रमिष्यामः ॥

Reference	Text *Devanāgarī*
1.9.15	वराहवंस्स्वतपसः । विद्युन्मंहसो धूपंयः । श्वापयो गृहमेधाश्चेत्येते । ये चेमेऽशिमिविद्विषः ॥
1.9.16	पर्जन्यास्सप्त पृथिवीमभिवंर्षन्ति । वृष्टिंभिरिति । एतयैव विभक्तिविंपरीताः । सप्तभिर्वातैरुदीरिताः । अमूँल्लोकानभिवंर्षन्ति । तेषांमेषा भवंति ॥
1.9.17	समानमेतदुदकम् । उच्चैत्यंवचाहंभिः । भूमिं पर्जन्या जिन्वंन्ति । दिवं जिन्वन्त्यग्नय इति ॥
1.9.18	यदक्षरं भूतकृतम् । विश्वे देवा उपासंते । महर्षिमस्य गोप्तारम् । जमदंग्निमकुर्वत ॥
1.9.19	जमदंग्निराप्यायते । छन्दोंभिश्चतुरुत्तरैः । राग्यस्सोमंस्य तृप्तासः । ब्रह्मणा वीर्यावता । शिवा नः प्रदिशो दिशः ॥
1.9.20	तच्छंयोरावृंणीमहे । गातुं यग्याय । गातुं यग्यपंतये । दैवीस्वस्तिरंस्तु नः । स्वस्तिर्मानुषेभ्यः । ऊर्ध्वं जिंगातु भेषजम् । शन्नो अस्तु द्विपदे । शं चतुंष्पदे ॥
1.9.21	सोमपा३ असोमपा३ इति निगदंव्याख्याताः ॥
1.10.1	सहस्रवृदियं भूमिः । परं व्योम सहस्रंवृत् । अश्विनां भुज्यू नासत्या । विश्वस्यं जगतस्पंती ॥
1.10.2	जाया भूमिः पंतिर्व्योम । मिथुनंन्ता अतुर्यथुः । पुत्रो बृहस्पंती रुद्रः । सरमा इति स्त्रीपुमम् ॥
1.10.3	शुक्रं वामन्यद्यंजतं वामन्यत् । विषुंरूपे अहंनी द्यौरिव स्थः । विश्वा हि माया अवथः स्वधावन्तो । भद्रा वां पूषणाविह रातिरंस्तु ॥
1.10.4	वासात्यौ चित्रौ जगंतो निधानौ । द्यावांभूमी चरथः सꣳसखायौ । तावश्विनां रासभाश्वा हवं मे । शुभस्पती आगतꣳसूर्यया सह ॥

Reference	**Text *Devanāgarī***
1.10.5	त्युग्रो॑ह भु॒ज्युम॑श्विनोदमे॒घे । र॒यिन्न कश्चि॑न्ममृ॒वां २ अवा॑हाः । तमू॑हथुर्नौ॒भिरा᳚त्म॒न्वती॑भिः । अ॒न्त॒रि॒क्ष॒प्रुद्भि॑र॒पोद॑काभिः ॥
1.10.6	ति॒स्रः क्षप॒स्त्रिर॒हा॑ऽति॒व्रज॑द्भिः । नास॑त्या भु॒ज्युमू॑हथुः पत॒ङ्गैः । स॒मु॒द्रस्य॒ धन्व॑न्ना॒र्द्रस्य॑ पा॒रे । त्रि॒भीरथै᳚श्श॒तप॑द्भिः षड॑श्वैः ॥
1.10.7	स॒वि॒ता॒रं॒ वित॑न्वन्तम् । अनु॑बध्नाति॒ शाम्ब॒रः । आप॑पूरुषम्ब॑रश्चैव । स॒विता॑ऽरेप॒सो॑ भव॒त् ॥
1.10.8	त्यँ सुतृप्तं विं॑दित्वैव । ब॒हु॒सो॑म॒ गि॒रं व॑शी । अन्वेति तुग्रो व॑क्रिया॒न्तम् । आय॑सूया॒न्थ्सोम॑तृप्सु॒षु ॥
1.10.9	स सं॑ग्रामस्त॒मो᳚द्योऽत्यो॒तः । वा॒चो गाः पिं॑पाति॒ तत् । स तद्गोभिस्स्त्वा᳚ऽत्येत्य॒न्ये । र॒क्षसा॑ऽनन्वि॒ताश्च॑ ये ॥
1.10.10	अन्वेति परि॑वृत्यास्तः । ए॒वमे॒तौ स्थो॑ अश्विना । ते ए॒ते द्युं॑ः पृ॒थि॒व्योः । अह॑र॒हर्गर्भ॑न्दधाथे ॥
1.10.11	तयो॑रेतौ व॒थ्सावं॑हो॒रात्रे । पृ॒थिव्या अह॑ः । दि॒वो रात्रिं॑ः । ता अवि॑सृष्टौ । दम्प॑ती ए॒व भ॑व॒तः ॥
1.10.12	तयो॑रेतौ व॒थ्सौ । अ॒ग्निश्चा॑दित्यश्च॑ । रा॒त्रेर्व॒थ्सः । श्वेत आ॑दि॒त्यः । अह्नो॒ऽग्निः । ता॒म्रो अ॑रु॒णः । ता अवि॑सृष्टौ । दम्प॑ती ए॒व भ॑वतः ॥
1.10.13	तयो॑रेतौ व॒थ्सौ । वृ॒त्रश्च॑ वैद्युतश्च॑ । अ॒ग्नेर्वृ॒त्रः । वै॒द्युतं॑ आदि॒त्यस्य॑ । ता अवि॑सृष्टौ । दम्प॑ती ए॒व भ॑व॒तः ॥
1.10.14	तयो॑रेतौ॒ व॒थ्सौ । उ॒ष्मा च॑ नीहा॒रश्च॑ । वृ॒त्रस्यो॒ष्मा । वै॒द्यु॒तस्य॑ नीहा॒रः । तौ ताव॒व प्रति॑पद्येते ॥
1.10.15	से यꣳरा॒त्रीं॑गर्भि॒णी॑ पु॒त्रेण॒ सं॑वसति । तस्या॒ वा ए॒तद॒ल्बण᳚म् । यद्रा॒त्रौ र॒श्मय॑ः । य॒था॒ गोर्ग॒र्भिण्या उ॒ल्बण᳚म् । ए॒वमे॒तस्या॑ उ॒ल्बण᳚म् ॥

Reference	Text *Devanāgarī*
1.10.16	प्रजयिष्णुः प्रजया च पशुभि॑श्च भवति । य ए॑वं वेद । एतमुद्यन्तमपिय॑न्तं चे॒ति । आदि॑त्यः पुण्य॑स्य व॒त्सः ॥
1.10.17	अथ पवि॑त्राङ्गि॒रसः ॥
1.11.1	प॒वित्र॑वन्तः परि॒वाजमास॑ते । पि॒तैषां᳚ प्र॒त्नो अ॒भिर॑क्षति व्र॒तम् । म॒हस्स॑मु॒द्रं वरु॑णस्ति॒रोद॑धे । धीरा॑ इच्छेकु॒र्धरु॑णेष्वा॒रभ᳚म् ॥
1.11.2	प॒वित्र॑ ते॒ वित॑तं॒ ब्रह्म॑णस्पते᳚ । प्र॒भुर्गात्रा॑णि॒ पर्ये॑षि वि॒श्वतः॑ । अत॑प्ततनू॒र्न तदा॒मो अ॑श्नुते । शृ॒ता॑स॒ इद्वह॑न्त॒स्तत्समा॑शत ॥
1.11.3	ब्र॒ह्मा दे॒वाना᳚म् ॥
1.11.4	अ॒स॑तस्स॒द्ये तत॑क्षुः । ऋष॑यस्सप्ता॒त्रिश्च॒ यत् । सर्वे॑ऽत्रयो अग॑स्त्य॒श्च । नक्ष॑त्रै॒श्शंकृ॑तोऽवस॑न् ॥
1.11.5	अथ॑ सवि॒तुः श्यावाश्व॒स्यावर्ति॑कामस्य ॥
1.11.6	अ॒मी य ऋक्षा॒ निहि॑तास उच्चा । नक्तं॒ दद॑श्रे॒ कुह॑चि॒द्दिवे॑युः । अद॑ब्धानि॒ वरु॑णस्य व्र॒तानि॑ । वि॒चा॒कश॑च्च॒न्द्रमा॒ नक्ष॑त्रमेति ॥
1.11.7	तत्स॑वि॒तुर्वरे᳚ण्यम् । भर्गो॑ दे॒वस्य॑ धीमहि । धियो॒ यो नः॑ प्रचो॒दया᳚त् ॥
1.11.8	तत्स॑वि॒तुर्वृ॑णीमहे । वयन्दे॒वस्य॒ भोज॑नम् । श्रेष्ठ॑ꣳसर्व॒धात॑मम् । तुरं॒ भग॑स्य धीमहि ॥
1.11.9	अपा॑गूहत सवि॒ता तृभीन्॑ । सर्वा᳚न्दि॒वो अन्ध॑सः । नक्त॒न्तान्य॑भवन्दृ॒शे । अस्थ्य॒स्था स॑म्भविष्यामः ॥
1.11.10	नाम॒ नामै॑व नाम॒ मे᳚ । नपुꣳस॑कं पुमाꣳस्त्र्य॑स्मि । स्थाव॑रोऽस्म्यथ॒ जङ्ग॑मः । य॒जेऽय॑क्षि॒ यष्टा॒हे च॑ ॥

Reference	Text *Devanāgarī*
1.11.11	मया॑ भू॒तान्य॑यक्षत । प॒शवो॑ मम॑ भू॒तानि॑ । अनू॒बन्ध्यो॑ऽस्म्य॒हं वि॒भुः ॥
1.11.12	स्त्रिय॑स्स॒तीः । ता उ॑ मे पु॒ꣳस आ॑हुः । पश्य॑दक्ष॒ण्वान्नविचे॑त॒दन्धः । क॒विर्यः पु॒त्रस्स इ॒मा चि॑केत । यस्ता वि॑जा॒नाथ्स॑वि॒तुः पि॒ता स॑त् ॥
1.11.13	अ॒न्धो म॑णि॒मं॑विन्दत् । तम॑नङ्गु॒लिराव॑यत् । अ॒ग्री॒वः प्र॒त्य॑मुञ्चत् । तम॑जि॒ह्वा अ॒सश्च॑त ॥
1.11.14	ऊ॒र्ध्वमू॑लम॒वाक्छा॑खम् । वृ॒क्षं यो॑ वेद॒ सम्प्र॑ति । न स जा॒तु जनः॑ श्रद्द॒ध्यात् । मृ॒त्युर्मा॑ मार॒यादि॑तिः ॥
1.11.15	ह॒सि॒तꣳरु॒दि॑तङ्गी॒तम् । वी॒णाप॑णवला॒सि॑तम् । मृ॒तञ्जी॒वं च॑ यत्कि॒ञ्चित् । अ॒ङ्गानि॑ स्ने॒व वि॒द्धि॑ तत् ॥
1.11.16	अ॒तृ॑ष्य॒न्गस्तृ॒ष्य॑ ध्यायत् । अ॒स्माज्जा॒ता मे॑ मि॒थू च॑रन् । पु॒त्रो निर्ऋ॑त्या वैदे॒हः । अ॒चेता॑ य॒श्च॒ चेत॑नः ॥
1.11.17	स॒ तं म॑णि॒मं॑विन्दत् । सो॑ऽग्री॒वः प्र॒त्य॑मुञ्चत् । सो॑ऽजि॒ह्वो अ॒सश्च॑त ॥
1.11.18	नैत॑मृ॒षिं वि॑दि॒त्वा न॒ग॑रं प्र॒विशे॑त् । य॑दि प्र॒विशे॑त् । मि॒थौ च॒रि॑त्वा प्र॒विशे॑त् । त॒थ्सम्भ॑व॒स्य व्र॒तम् ॥
1.11.19	आत॑म॒ग्ने र॒थन्ति॑ष्ठ । एका॑श्वमेक॒योज॑नम् । एक॒चक्र॑मेक॒धुरम् । वा॒तध्रा॑जि॒गति॑ वि॒भो ॥
1.11.20	न रि॒ष्यति॑ न व्यथते । ना॒स्याक्षो॑ या॒तु स॒ज्ज॑ति । य॒च्छ्वे॒ता॒न्रोहि॑ता ꣳश्चा॒ग्नेः । र॒थे यु॑क्त्वाऽधि॒तिष्ठ॑ति ॥
1.11.21	एक॑या च द॒शभि॑श्च स्वभूते । द्वाभ्या॑मि॒ष्टये॑ वि॒ꣳश॒त्या च॑ । ति॒सृभि॑श्च वहसे त्रि॒ꣳश॒ता च॑ । नियु॒द्भिर्वा॑यवि॒हिता॑ वि॒मु॒ञ्च ॥

Reference	Text *Devanāgarī*
1.12.1	आतनुष्व प्रतनुष्व । उद्धमाऽऽधम सन्धम । आदित्ये चन्द्रवर्णानाम् । गर्भमाधेहि यः पुमान् ॥
1.12.2	इतस्सिक्त३ सूर्यगतम् । चन्द्रमसे रसङ्कृधि । वारादञ्जनयाग्रेऽग्निम् । य एको रुद्र उच्यते ॥
1.12.3	असंख्यातास्सहस्राणि । स्मर्यते न च दृश्यते । एवमेतन्निबोधत ॥
1.12.4	आमन्द्रैरिन्द्र हरिभिः । याहि मयूररोमभिः । मा त्वा केचिन्नियेमुरिन्न पाशिनः । दधन्वेव ता इहि ॥
1.12.5	मा मन्द्रैरिन्द्र हरिभिः । यामि मयूररोमभिः । मा मा केचिन्नियेमुरिन्न पाशिनः । निधन्वेव तां २ इमि ॥
1.12.6	अणुभिश्च महद्भिश्च । निघृष्वैरसमायुतैः । कालैर्हरित्वमापन्नैः । इन्द्रायाहि सहस्रयुक् ॥
1.12.7	अग्निर्विभ्राष्टिवसनः । वायुश्श्वेतसिकद्रुकः । संवथ्सरो विषूवर्णैः । नित्यास्तेऽनुचरास्तव ॥
1.12.8	सुब्रह्मण्यो३ सुब्रह्मण्यो३ सुब्रह्मण्योम् । इन्द्रागच्छ हरिव आगच्छ मेधातिथेः । मेष वृषणश्वस्य मेने । गौरावस्कन्दिन्नहल्यायै जार । कौशिकब्राह्मण गौतमब्रुवाण ॥
1.12.9	अरुणाश्वा इहागताः । वसवः पृथिविक्षितः । अष्टौ दिग्वाससोऽग्नयः । अग्निश्च जातवेदाश्चेत्येते ॥
1.12.10	तम्राश्वाँस्ताम्ररथाः । ताम्रवर्णास्तथाऽसिताः । दण्दहस्ताः खादग्दतः । इतो रुद्राः पराङ्गताः । उक्तग्गस्थानं प्रमाणञ्च पुर इत ॥
1.12.11	बृहस्पतिश्च सविता च । विश्वरूपैरिहागताम् । रथेनौदकवर्त्मना । अप्सुषा इति तद्द्वयोः ॥
1.12.12	उक्तो वेषो वासा३सि च । कालावयवानामितः प्रतीच्या । वासात्या इत्यश्विनोः । कोऽन्तरिक्षे शब्दङ्करोतीति । वासिष्ठो रौहिणो मी मं३साञ्चक्रे । तस्यैषा भवति ॥

Reference	Text *Devanāgarī*
1.12.13	वाश्रेवं विद्युदितिं ॥
1.12.14	ब्रह्मंण उदरंणमसि । ब्रह्मंण उदीरणंमसि । ब्रह्मंण आस्तरंणमसि । ब्रह्मंण उपस्तरंणमसि ॥
1.13.1	अष्टयोनीमष्टपुत्राम् अष्टपंत्नीमिमां महीम् । अहं वेद न में मृत्युः । नचामृंत्युरघाऽहंरत् ॥
1.13.2	अष्टयोंन्यष्टपुंत्रम् । अष्टपंदिदमन्तरिंक्षम् । अहं वेद न में मृत्युः । नचामृंत्युरघाऽहंरत् । अष्टयोनीमष्टपुत्राम् । अष्टपंत्नीममून्दिवंम् अहं वेद न में मृत्युः । नचामृंत्युरघाऽऽहंरत् ॥
1.13.3	सुत्रामाणं महीमूषु ॥
1.13.4	अदिंतिर्माता स पिता स पुत्रः । विश्वें देवा अदिंतिः पञ्चजनाः । अदिंतिर्जातमदिंतिर्जनिंत्वम् ॥
1.13.5	अष्टौ पुत्रासो अदिंतेः । ये जातास्तन्वंः परिं । देवां २ उपंप्रैत्सप्तभिंः । परा मार्ताण्डमास्यत् ॥
1.13.6	सप्तभिंः पुत्रैरदिंतिः । उपप्रैत्पूर्व्यं युगम् । प्रजायै मृत्यवे तंत् । परा मार्ताण्डमाभरदितिं ॥
1.13.7	ताननुक्रंमिष्यामः ॥
1.13.8	मित्रश्च वरुंणश्च । धाता चार्यमा चं । अꣳशश्च भगंश्च । इन्द्रश्च विवस्वाग्श्चेत्येते ॥
1.13.9	हिरण्यगर्भो हꣳसश्शुंचिषत् । ब्रह्मंग्यानं तदित्पदमितिं ॥
1.13.10	गर्भः प्रांजापत्यः । अथ पुरुंषः सप्तपुरुंषः ॥
1.14.1	योऽसौं तपन्नुदेतिं । स सर्वेषां भूतानां प्राणानादायोदेतिं । मा में प्रजया मा पंशूनाम् । मा मम प्राणानादायोदगाः ॥

Reference	Text *Devanāgarī*
1.14.2	असौ योऽस्तमेति । स सर्वेषां भूतानां प्राणानादायास्तमेति । मा मे प्रजया मा पशूनाम् । मा मम प्राणानादायाऽस्तङ्गाः ॥
1.14.3	असौ य आपूर्यति । स सर्वेषां भूतानां प्राणैरापूर्यति । मा मे प्रजया मा पशूनाम् । मा मम प्राणैरापूरिष्ठाः ॥
1.14.4	असौ योऽपक्षीयति । स सर्वेषां भूतानां प्राणैरपक्षीयति । मा मे प्रजया मा पशूनाम् । मा मम प्राणैरपक्षेष्ठाः ॥
1.14.5	अमूनि नक्षत्राणि । सर्वेषां भूतानां प्राणैरपप्रसर्पन्ति चोथ्सर्पन्ति च । मा मे प्रजया मा पशूनाम् । मा मम प्राणैरपप्रसृपत मोथ्सृपत ॥
1.14.6	इमे मासाश्चार्धमासाश्च । सर्वेषां भूतानां प्राणैरपप्रसर्पन्ति चोथ्सर्पन्ति च । मा मे प्रजया मा पशूनाम् । मा मम प्राणैरपप्रसृपत मोथ्सृपत । इम ऋतवः । सर्वेषां भूतानां प्राणैरपप्रसर्पन्ति चोथ्सर्पन्ति च । मा मे प्रजया मा पशूनाम् । मा मम प्राणैरपप्रसृपत मोथ्सृपत ॥
1.14.7	अयꣳ संवथ्सरः । सर्वेषां भूतानां प्राणैरपप्रसर्पति चोथ्सर्पति च । मा मे प्रजया मा पशूनाम् । मा मम प्राणैरपप्रसृप मोथ्सृप ॥
1.14.8	इदमहः । सर्वेषां भूतानां प्राणैरपप्रसर्पति चोथ्सर्पति च । मा मे प्रजया मा पशूनाम् । मा मम प्राणैरपप्रसृप मोथ्सृप । इयँरात्रिः । सर्वेषां भूतानां प्राणैरपप्रसर्पति चोथ्सर्पति च । मा मे प्रजया मा पशूनाम् । मा मम प्राणैरपप्रसृप मोथ्सृप ॥
1.14.9	ओं भुर्भुवस्स्वः ॥

Reference	Text *Devanāgarī*
1.14.10	एतद्वो मिथुनं मानो मिथुन ॰ रीढ्वम् ॥
1.15.1	अथादित्यस्याष्टपुरुषस्य ॥
1.15.2	वसूनामादित्यानाग्स्थाने स्वतेजसा भानि ॥
1.15.3	रुद्रणामादित्यानाग्स्थाने स्वतेजसा भानि ॥
1.15.4	आदित्यानामादित्यानाग्स्थाने स्वतेजसा भानि ॥
1.15.5	सता ॰ सत्यानाम् । आदित्यानाग्स्थाने स्वतेजसा भानि ॥
1.15.6	अभिधून्वतांमभिघ्नताम् । वातवंतां मरुताम् । आदित्यानाग्स्थाने स्वतेजसा भानि ॥
1.15.7	ऋभूणामादित्यानाग्स्थाने स्वतेजसा भानि ॥
1.15.8	विश्वेषान्देवानाम् । आदित्यानाग्स्थाने स्वतेजसा भानि ॥
1.15.9	संवत्सरस्य सवितुः । आदित्यस्य स्थाने स्वतेजसा भानि ॥
1.15.10	ओं भुर्भुवस्स्वः । रश्मयो वो मिथुनं मा नो मिथुन ॰ रीढ्वम् ॥
1.16.1	आरोगस्य स्थाने स्वतेजसा भानि । भ्राजस्य स्थाने स्वतेजसा भानि । पटरस्य स्थाने स्वतेजसा भानि । पतङ्गस्य स्थाने स्वतेजसा भानि । स्वर्णरस्य स्थाने स्वतेजसा भानि । ज्योतिषीमंतस्य स्थाने स्वतेजसा भानि । विभासस्य स्थाने स्वतेजसा भानि । कश्यपस्य स्थाने स्वतेजसा भानि । ओं भुर्भुवस्स्वः । आपो वो मिथुनं मा नो मिथुन ॰ रीढ्वम् ॥

Reference	Text *Devanāgarī*
1.17.1	अथ वायोरेकादशपुरुषस्यैकादशस्त्रीकस्य । प्रभ्राजमानानाꣳ रुद्राणाग्स्थाने स्वतेजसा भानि । व्यवदातानाꣳ रुद्राणाग्स्थाने स्वतेजसा भानि । वासुकिवैद्युतानाꣳ रुद्राणाग् स्थाने स्वतेजसा भानि । रजतानाꣳ रुद्राणाग् स्थाने स्वतेजसा भानि । परुषाणाꣳ रुद्राणाग् स्थाने स्वतेजसा भानि । श्यामानाꣳ रुद्राणाग् स्थाने स्वतेजसा भानि । कपिलानाꣳ रुद्राणाग् स्थाने स्वतेजसा भानि । अतिलोहितानाꣳ रुद्राणाग् स्थाने स्वतेजसा भानि । ऊर्ध्वानाꣳ रुद्राणाग् स्थाने स्वतेजसा भानि । अवपतन्तानाꣳ रुद्राणाग् स्थाने स्वतेजसा भानि । वैद्युतानाꣳ रुद्राणाग् स्थाने स्वतेजसा भानि । प्रभ्राजमानीनाꣳ रुद्राणीनाग् स्थाने स्वतेजसा भानि । व्यवदातीनाꣳ रुद्राणीनाग् स्थाने स्वतेजसा भानि । वासुकिवैद्युतीनाꣳ रुद्राणीनाग् स्थाने स्वतेजसा भानि । रजतानाꣳ रुद्राणीनाग् स्थाने स्वतेजसा भानि । परुषाणाꣳ रुद्राणीनाग् स्थाने स्वतेजसा भानि । श्यामानाꣳ रुद्राणीनाग् स्थाने स्वतेजसा भानि । कपिलानाꣳ रुद्राणीनाग् स्थाने स्वतेजसा भानि । अतिलोहितानाꣳ रुद्राणीनाग् स्थाने स्वतेजसा भानि । ऊर्ध्वानाꣳ रुद्राणीनाग् स्थाने स्वतेजसा भानि । अवपतन्तीनाꣳ रुद्राणीनाग् स्थाने स्वतेजसा भानि । वैद्युतीनाꣳ रुद्राणीनाग् स्थाने स्वतेजसा भानि । ओं भुर्भुवस्स्वः । रूपाणि वो मिथुनं म नो मिथुनꣳरीढ्वम् ॥
1.18.1	अग्नेः पूर्वदिशस्य स्थाने स्वतेजसा भानि । जातवेदस्य उपदिशस्य स्थाने स्वतेजसा भानि । सहोजसो दक्षिणदिशस्य स्थाने स्वतेजसा भानि । अजिराप्रभव उपदिशस्य स्थाने स्वतेजसा भानि । वैश्वानरस्यापरदिशस्य स्थाने स्वतेजसा भानि । नर्यापस उपदिशस्य स्थाने स्वतेजसा भानि । पङ्क्तिराधस उदग्दिशस्य स्थाने स्वतेजसा भानि । विसर्पिण उपदिशस्य स्थाने स्वतेजसा भानि । ओं भुर्भुवस्स्वः । दिशो वो मिथुनं मा नो मिथुनँरीढ्वम् ॥
1.19.1	दक्षिणपूर्वस्यान्दिशि विसर्पी नरकः । तस्मान्नः परिपाहि ॥
1.19.2	दक्षिणाऽपरस्यान्दिश्यविसर्पी नरकः । तस्मान्नः परिपाहि ॥
1.19.3	उत्तरपूर्वस्यान्दिशि विषादी नरकः । तस्मान्नः परिपाहि ॥

Reference	**Text *Devanāgarī***
1.19.4	उत्तरपरस्यान्दिश्यविषादी नरकः । तस्मान्नः परिपाहि । आ यस्मिन्स्थप्त वासावा इन्द्रियाणि शतक्रतवित्येते ॥
1.20.1	इन्द्रघोषा वो वसुभिः पुरस्तादपुदधताम् । मनोजवसो वः पितृभिर्दक्षिणत उपदधताम् । प्रचेता वो रुद्रैः पश्चादपुदधताम् । विश्वकर्मा व आदित्यैरुत्तरत उपदधताम् । त्वष्टा वो रूपैरुपरिष्टादपुदधताम् । संज्ञानं वः पश्चादिति ॥
1.20.2	आदित्यस्सर्वोऽग्निः पृथिव्याम् । वायुरन्तरिक्षे । सूर्यो दिवि । चन्द्रमा दिक्षु । नक्षत्राणि स्वलोके ॥
1.20.3	एवा ह्येव । एवा ह्यग्ने । एवा हि वायो । एवा हीन्द्र । एवा हि पूषन् । एवा हि देवाः ॥
1.21.1	आपमापामपः सर्वाः । अस्मादस्मादितोऽमुतः । अग्निर्वायुश्च सूर्यश्च । सह संचस्करर्द्धिया । वाय्वश्वा रश्मिपतयः । मरीच्यात्मानो अद्रुहः । देवीर्भुवनसूवरीः । पुत्रवत्त्वाय मे सुत । महानाम्नीर्महामानाः । महसो महसस्स्वः । देवीः पर्जन्यसूवरीः । पुत्रवत्त्वाय मे सुत । अपाऽश्न्युष्णिमपारक्षः । अपाऽश्न्युष्णिमपारघम् । अपाघ्रामपचाऽवर्तिम् । अपदेवीरितो हित । वज्रन्देवीरजीताग्श्च । भुवनन्देवसूवरीः । आदित्यानदितिन्देवीम् । योनिनोर्ध्वमुदीषत । भद्रं कर्णेभिः श्रृणुयाम देवाः । भद्रं पश्येमाक्षभिर्यजत्राः । स्थिरैरङ्गैस्तुष्टुवाꣳ संस्तनूभिः । व्यशेम देवहितं यदायुः । स्वस्ति न इन्द्रो वृध्दश्रवाः । स्वस्ति नः पूषा विश्ववेदाः । स्वस्तिनस्ताक्ष्र्यो अरिष्टनेमिः । स्वस्ति नो बृहस्पतिर्दधातु । केतवो अरुणासश्च । ऋषयो वातरशनाः । प्रतिष्ठाꣳशतधा हि । समाहितासो ऽसहस्रधायसम् । शिवा नश्शन्तमा भवन्तु । दिव्या आप ओषधयः । सुमृडीका सरस्वति । मा ते व्योम संदृशि ॥
1.22.1	योऽपां पुष्पं वेद । पुष्पवान्प्रजावान्पशुमान्भवति । चन्द्रमा वा अपां पुष्पम् । पुष्पवान्प्रजावान्पशुमान्भवति । य एवं वेद ॥

Reference	Text *Devanāgarī*
1.22.2	योऽपामायतनं वेद । आयतनवान्भवति । अग्निर्वा अपामायतनम् । आयतनवान्भवति । योऽग्नेरायतनं वेद । आयतनवान्भवति । आपो वा अग्नेरायतनम् । आयतनवान्भवति । य एवं वेद ॥
1.22.3	योऽपामायतनं वेद । आयतनवान्भवति । वायुर्वा अपामायतनम् । आयतनवान्भवति । यो वायोरायतनं वेद । आयतनवान्भवति । आपो वै वायोरायतनम् । आयतनवान्भवति । य एवं वेद ॥
1.22.4	योऽपामायतनं वेद । आयतनवान्भवति । असौ वै तपन्नपामायतनम् । आयतनवान्भवति । योऽमुष्य तपत आयतनं वेद । आयतनवान्भवति । आपो वा अमुष्य तपत आयतनम् । आयतनवान्भवति । य एवं वेद ॥
1.22.5	योऽपामायतनं वेद । आयतनवान्भवति । चन्द्रमा वा अपामायतनम् । आयतनवान्भवति । यश्चन्द्रमस आयतनं वेद । आयतनवान्भवति । आपो वै चन्द्रमस आयतनम् । आयतनवान्भवति य एवं वेद ॥
1.22.6	योऽपामायतनं वेद । आयतनवान्भवति । नक्षत्राणि वा अपामायतनम् । आयतनवान्भवति । यो नक्षत्राणामायतनं वेद । आयतनवान्भवति । आपो वै नक्षत्राणामायतनम् । आयतनवान्भवति । य एवं वेद ॥
1.22.7	योऽपामायतनं वेद । आयतनवान्भवति । पर्जन्यो वा अपामायतनम् । आयतनवान्भवति । यः पर्जन्यस्याऽऽयतनं वेद । आयतनवान्भवति । आपो वै पर्जन्यस्याऽऽयतनम् । आयतनवान्भवति । य एवं वेद ॥

Reference	**Text *Devanāgarī***
1.22.8	योऽपामायतनं वेद आयतनवान्भवति । संवथ्सरो वा अपामायतनम् । आयतनवान्भवति । यस्संवथ्सरस्यायतनं वेद । आयतनवान्भवति । आपो वै संवथ्सरस्याऽऽयतनम् । आयतनवान्भवति । य एवं वेद ॥
1.22.9	योऽप्सु नावं प्रतिष्ठितां वेद । प्रत्येव तिष्ठति इमे वै लोका अप्सु प्रतिष्ठिताः । तदेषाऽभ्यनूक्ता ॥
1.22.10	अपाꣳरसमुदयꣳसन्न् । सूर्ये शुक्रँसमाभृतम् । अपाꣳरसस्य यो रसः । तं वो गृह्णाम्युत्तममिति ॥
1.22.11	इमे वै लोका अपाँरसः । तेऽमुष्मिन्नादित्ये समाभृताः ॥
1.22.12	जानुद्रघ्नीमुत्तरवेदीङ्खात्वा । अपां पूरयित्वा गुल्फदघ्नम् । पुष्करपर्णैः पुष्करदण्डैः पुष्करैश्च सग्ग्स्तीर्य । तस्मिन्विहायसे । अग्निं प्रणीयोपसमाधाय ॥
1.22.13	ब्रह्मवादिनो वदन्ति । कस्मात्प्रणीतेऽयमग्निश्चीयते । साप्रणीतेऽयमप्सु ह्ययंश्चीयते । असौ भुवनेऽप्यनाहिताग्निरेताः । तमभित एता अबीष्टका उपदधाति ॥
1.22.14	अग्निहोत्रे दर्शपूर्णमासयोः । पशुबन्धे चातुर्मास्येषु । अथो आहुः । सर्वेषु यग्यक्रतुष्विति ॥
1.22.15	एतद्ध स्म वा आहुश्शण्डिलाः । कमग्निश्चिनुते । सत्रियमग्निश्चिन्वानः । संवथ्सरं प्रत्यक्षेण ॥
1.22.16	कमग्निश्चिनुते । सावित्रमग्निश्चिन्वानः । अमुमादित्यं प्रत्यक्षेण ॥
1.22.17	कमग्निश्चिनुते । नाचिकेतमग्निश्चिन्वानः । प्राणान्प्रत्यक्षेण ॥
1.22.18	कमग्निश्चिनुते । चातुर्होत्रियमग्निश्चिन्वानः । ब्रह्म न्प्रत्यक्षेण ॥
1.22.19	कमग्निश्चिनुते । वैश्वसृजमग्निश्चिन्वानः । शरीरं न्प्रत्यक्षेण ॥

Reference	Text *Devanāgarī*
1.22.20	कमग्निश्चिनुते । उपानुवाक्यमाशुमग्निश्चिन्वानः । इमाँल्लोकान्प्रत्यक्षेण ॥
1.22.21	कमग्निश्चिनुते । इममारुणकेतुकमग्निश्चिन्वानः इति । य एवासौ । इतश्चामुतश्चाऽव्यतीपाती । तमिति ॥
1.22.22	योऽग्नेर्मिथूया वेद । मिथुनवान्भवति । य एवं वेद ॥
1.23.1	आपो वा इदमासन्त्सलिलमेव । स प्रजापतिरेकः पुष्करपर्णे समभवत् । तस्यान्तर्मनसि कामस्समवर्तत । इदꣳ सृजेयमिति । तस्माद्यत्पुरुषो मनसाऽभिगच्छति । तद्वाचा वदति । तत्कर्मणा करोति । तदेषाऽभ्यनूक्ता ॥
1.23.2	कामस्तदग्रे समवर्तताधि । मनसो रेतः प्रथमं यदासीत् ॥ सतो बन्धुमसति निरविन्दन् । हृदि प्रतीष्या कवयो मनीषेति ॥
1.23.3	उपैनन्तदपुनमति । यत्कामो भवति । य एवं वेद ॥
1.23.4	स तपोऽतप्यत । स तपस्तप्त्वा । शरीरमधूनुत । तस्य यन्माꣳसमासीत् । ततोऽरुणाः केतवो वातरशना ऋषय उदतिष्ठन्न् । ये नखाः । ते वैखानसाः । ये वालाः । ते वालखिल्याः । यो रसः । सोऽपाम् ॥
1.23.5	अन्तरतः कूर्मं भूतꣳसर्पन्तम् । तमब्रवीत् । मम वैत्वङ्माꣳसा । समभूत् । नेत्यब्रवीत् । पूर्वमेवाहमिहासमिति । तत्पुरुषस्य पुरुषत्वम् । स सहस्रशीर्षा पुरुषः । सहस्राक्षस्सहस्रपात् । भूत्वोदतिष्ठत् । तमब्रवीत् । त्वं वै पूर्वꣳ समभूः । त्वमिदं पूर्वः कुरुष्वेति ॥
1.23.6	स इत आदायापः । अञ्जलिना पुरस्तादुपादधात् । एवाह्येवेति । तत आदित्य उदतिष्ठत् । सा प्राची दिक् ॥

Reference	Text *Devanāgarī*
1.23.7	अथा॑रु॒णः के॒तुर्द॑क्षिण॒त उपा॑दधात् । ए॒वाह्य॒ग्न इति॑ । त॒तो॒ वा अ॒ग्निरु॒द॑तिष्ठत् । सा द॑क्षि॒णा दिक् । अथा॑रु॒णः के॒तुः प॒श्चाद॒पुपा॑दधात् । ए॒वाहि वा॑यो॒ इति॑ । त॒तो वा॒युरु॒द॑तिष्ठत् । सा प्र॒तीची॒ दिक् । अथा॑रु॒णः के॒तुरु॑त्तर॒त उपा॑दधात् । ए॒वाही॒न्द्रेति॑ । त॒तो॒ वा इन्द्र॒ उद॑तिष्ठत् । सोदी॑ची॒ दिक् । अथा॑रु॒णः के॒तुर्म॒ध्य उपा॑दधात् । ए॒वाहि॒ पूष॒न्निति॑ । त॒तो॒ वै पू॒षोद॑तिष्ठत् । सेयन्दिक् । अथा॑रु॒णः के॒तुरु॒परि॑ष्टाद॒पुपा॑दधात् । ए॒वाहि दे॒वा इति॑ । त॒तो दे॑वम॒नु॒ष्याः॒ पि॒तरः॑ । ग॒न्ध॒र्वा॒प्स॒र॒स॒श्चोद॑तिष्टन् । सोर्ध्वा दिक् ॥
1.23.8	या वि॒प्रुषो॑ वि॒परा॑पतन्न् । ताभ्यो॒ऽसु॑रा रक्षाꣳसि पि॒शा॒चाश्चोद॑तिष्ठन्न् । तस्मा॒त्ते परा॑भवन्न् । वि॒प्रुड्भ्यो॒ हि ते सम॑भवन्न् ॥
1.23.9	त॒दे॒षाऽभ्य॑नू॒क्ता ॥
1.23.10	आपो॑ ह॒ यद्बृ॑ह॒तीर्गर्भ॒मायन्न् । दक्ष॒न्दधा॑ना ज॒नय॑न्तीस्स्व॒यं॑भुम् । तत॑ इ॒मेऽद्ध्य॑सृज्यन्त॒ सर्गाः॑ । अ॒द्भ्यो॒ वा इ॒दꣳसम॑भूत् । तस्मा॑दि॒दꣳसर्व॒ं ब्रह्म॑ स्व॒यं॒भ्विति॑ ॥
1.23.11	तस्मा॑दि॒दꣳसर्व॒ꣳशिथि॒लमि॒वाद्ध्रु॒वमि॑वाभवत् ॥
1.23.12	प्र॒जाप॑तिर्वाव॒ तत् । आ॒त्मना॒ऽऽत्मान॑ं वि॒धाय॑ । तदे॒वानु॒प्रावि॑शत् ॥
1.23.13	त॒दे॒षाऽभ्य॑नू॒क्ता ॥
1.23.14	वि॒धाय॑ लो॒कान्वि॒धाय॑ भू॒तानि॑ । वि॒धाय॒ सर्वाः॑ प्र॒दिशो॒ दिश॑श्च । प्र॒जाप॑तिः प्रथम॒जा ऋ॒तस्य॑ । आ॒त्मना॒ऽऽत्मान॑म॒भिसंवि॑वे॒शेति॑ ॥
1.23.15	सर्व॑मे॒वेदमा॒प्त्वा । सर्व॑मव॒रुद्ध्य॑ । तदे॒वानु॒प्रवि॑शति । य ए॒वं वेद॑ ॥
1.24.1	चतु॑ष्टय्य आपो॑ गृह्णाति । च॒त्वारि॒ वा अपाꣳरू॒पाणि॑ । मे॒घो वि॒द्युत् । स्त॒न॒यि॒त्नुर्वृ॒ष्टिः । ता॒न्ये॒वाव॑रुन्धे ॥

Reference	Text *Devanāgarī*
1.24.2	आतपति वर्ष्या गृह्णाति । ताः पुरस्तादुपदधाति । एता वै ब्रह्मवर्चस्या आपः । मुखत एव ब्रह्मवर्चसमवरुन्धे । तस्मान्मुखतो ब्रह्मवर्चसितरः ॥
1.24.3	कूप्या गृह्णाति । ता दक्षिणत उपदधाति । एता वै तेजस्विनीरापः । तेज एवास्य दक्षिणतो दधाति । तस्माद्दक्षिणोऽर्धस्तेजस्वितरः ॥
1.24.4	स्थावरा गृह्णाति । ताः पश्चादुपदधाति । प्रतिष्ठिता वै स्थावराः । पश्चादेव प्रतितिष्ठति ॥
1.24.5	वहन्तीर्गृह्णाति । ता उत्तरत उपदधाति । ओजसा वा एता वहन्तीरिवोद्गतीरिव आकूजतीरिव धावन्तीः । ओज एवास्योत्तरतो दधाति । तस्मादुत्तरोऽर्ध ओजस्वितरः ॥
1.24.6	संभार्या गृह्णाति । ता मध्य उपदधाति । इयं वै संभार्याः । अस्यामेव प्रतितिष्ठति ॥
1.24.7	पल्वल्या गृह्णाति । ता उपरिष्टादुपदधाति । असौ वै पल्वल्याः । अमुष्यामेव प्रतितिष्ठति ॥
1.24.8	दिक्षूपदधाति । दिक्षु वा आपः । अन्नं वा आपः । अद्भ्यो चा अन्नञ्जायते । यदेवाद्भ्योऽन्नञ्जायते । तदवरुन्धे ॥
1.24.9	तं वा एतमरुणाः केतवो वातरशना ऋषयोऽचिन्वन् । तस्मादारुणकेतुकः ॥
1.24.10	तदेषाऽभ्यनूक्ता ॥
1.24.11	केतवो अरुणासश्च । ऋषयो वातरशनाः । प्रतिष्ठाꣳ शतधाहि । समाहितासो सहस्रधायसमिति ॥
1.24.12	शतशश्चैव सहस्रशश्च प्रतितिष्ठति । य एतमग्निञ्चिनुते । य उचैनमेवं वेद ॥
1.25.1	जानुदघ्नीमुत्तरवेदीङ्खात्वा । अपां पूरयति । आपꣳ सर्वत्वाय ॥

Reference	Text *Devanāgarī*
1.25.2	पुष्करपर्णꣳरुक्मं पुरुषमित्युपदधाति । तपो वै पुष्करपर्णम् । सत्यꣳरुक्मः । अमृतं पुरुषः । एतावद्वावास्ति । यावदेतत् । यावदेवास्ति । तदवरुन्धे ॥
1.25.3	कूर्ममुपदधाति । अपामेव मेधमवरुन्धे । अथो स्वर्गस्य लोकस्य समष्ट्यै ॥
1.25.4	आपमापामपस्सर्वाः । अस्मादस्मादितोऽमुतः । अग्निर्वायुश्च सूर्यश्च । सहसंचस्करर्द्धिया इति । वाय्वश्वा रश्मिपतयः ॥
1.25.5	लोक पृणच्छिद्रं पृण । यास्तिस्त्रः परमजाः ॥
1.25.6	इन्द्रघोषा वो वसुभिरेवाह्येवेति ॥
1.25.7	पञ्चचितय उपदधाति । पाङ्क्तोऽग्निः । यावनेवाग्निः । तञ्चिनुते ॥
1.25.8	लोकपृणया द्वितीयामुपदधाति । पञ्चपदा वै विराट । तस्या वा इयं पादः । अन्तरिक्षं पादः । द्यौः पादः । दिशः पादः । परोरजाः पादः ॥
1.25.9	विराज्येव प्रतितिष्ठति । य एतमग्निश्चिनुते । य उ चैनमेवं वेद ॥
1.26.1	अग्निं प्रणीयोपसमाधाय । तमभित एता अबीष्टका उपदधाति । अग्निहोत्रे दर्शपूर्णमासयोः । पशुबन्धे चातुर्मास्येषु । अथो आहुः । सर्वेषु यग्यक्रतुष्विति ॥
1.26.2	अथ हस्माहारुणस्स्वायंभुवः । सावित्रस्सर्वोऽग्निरित्यननुषङ्गं मन्यामहे । नाना वा एतेषा वीर्याणि ॥

Reference	Text *Devanāgarī*
1.26.3	कमग्निश्चिनुते । सत्रियमग्निश्चिन्वानः । कमग्निश्चिनुते । सावित्रमग्निश्चिन्वानः । कमग्निश्चिनुते । नाचिकेतमग्निश्चिन्वानः । कमग्निश्चिनुते । चातुर्होत्रियमग्निश्चिन्वानः । कमग्निश्चिनुते । वैश्वसृजमग्निश्चिन्वानः । कमग्निश्चिनुते । उपानुवाक्यमाशुमग्निश्चिन्वानः । कमग्निश्चिनुते । इममारुणकेतुकमग्निश्चिन्वान इति ॥
1.26.4	वृषा वा अग्निः । वृषाणो सग्ग्स्फालयेत् । हन्येतास्य यग्यः । तस्मान्नानुषज्यः ॥
1.26.5	सोत्तरवेदिषु क्रतुषु चिन्वीत । उत्तरवेद्याँह्यग्निश्चीयते ॥
1.26.6	प्रजाकामश्चिन्वीत । प्राजापत्यो वा एषोऽग्निः । प्राजापत्याः प्रजाः । प्रजावान्भवति । य एवं वेद ॥
1.26.7	पशुकामश्चिन्वीत । संग्यानं वा एतत्पशूनाम् । यदापः । पशूनामेव संग्यानेऽग्निश्चिनुते । पशुमान्भवति । य एवं वेद ॥
1.26.8	वृष्टिकामश्चिन्वीत । आपो वै वृष्टिः । पर्जन्यो वर्षुको भवति । य एवं वेद ॥
1.26.9	आमयावी चिन्वीत । आपो वै भेषजम् । भेषजमेवास्मै करोति । सर्वमायुरेति ॥
1.26.10	अभिचरग्ग्श्चिन्वीत । वज्रो वै आपः । वज्रमेव भ्रातृव्येभ्यः प्रहरति । स्तृणुत एनम् ॥
1.26.11	तेजस्कामो यशस्कामः । ब्रह्मवर्चसकामस्स्वर्गकामश्चिन्वीत । एतावद्वा वास्ति । यावदेतत् । यावदेवास्ति । तदवरुन्धे ॥
1.26.12	तस्यैतद्व्रतम् । वर्षति न धावेत् अमृतं वा आपः । अमृतस्यानन्तरित्यै ॥

Reference	Text *Devanāgarī*
1.26.13	नाप्सु मूत्रंपुरीषङ्कुर्यात् । न निष्ठींवेत् । न विवसंनस्स्नायात् । गुह्यो वा एषोऽग्निः । एतस्याग्नेरनंतिदाहाय ॥
1.26.14	न पुंष्करपर्णानि हिरंण्यं वाऽधितिष्ठेत् एतस्याग्नेरनंभ्यारोहाय ॥
1.26.15	न कूर्मस्याश्नीयात् । नोदकस्याघातुंकान्येनंमोदकानिं भवन्ति । अघातुंका आपंः । य एतमग्निञ्चिनुते । य उंचैनमेवं वेद ॥
1.27.1	इमानुंकं भुवंना सीषधेम । इन्द्रंश्च विश्वें च देवाः ॥
1.27.2	यज्ञश्चं नस्तन्वंश्च प्रजाश्चं । आदित्यैरिन्द्रंस्सह सींषधातु ॥
1.27.3	आदित्यैरिन्द्रस्सगंणो मरुद्भिः । अस्माकं भूत्ववि़ता तनूनाम् ॥
1.27.4	आप्लंवस्व प्रप्लंवस्व । आण्डी भंव ज मा मुहुः । सुखादीन्दुःखनिधनाम् । प्रतिंमुञ्चस्व स्वां पुरम् ॥
1.27.5	मरींचयस्स्वायंभुवाः । ये शंरीराण्यंकल्पयन् । ते ते देहङ्कल्पयन्तु । मा चं ते ख्या स्मं तीरिषत् ॥
1.27.6	उत्तिंष्ठत मा स्वंप्त । अग्निमिंच्छध्वं भारंताः । राज्ञस्सोमंस्य तृप्तासंः । सूर्येण सयुजोषसः ॥
1.27.7	युवां सुवासाः । अष्टाचंक्रा नवंद्वारा ॥
1.27.8	देवानां पूरंयोध्या । तस्याꣳहिरण्मंयः कोशः । स्वर्गो लोको ज्योतिषाऽऽवृतः ॥
1.27.9	यो वै तां ब्रह्मंणो वेद । अमृतेनाऽऽवृतां पुरीम् । तस्मै ब्रह्म चं ब्रह्मा च । आयुः कीर्तिं प्रजान्ददुः ॥

Reference	Text *Devanāgarī*
1.27.10	विभ्राजमानाꣳ हरिणीम् । यशसा संपरीवृताम् । पुरꣳ हिरण्मयीं ब्रह्मा विवेशाऽपराजिता ॥
1.27.11	पराङेत्यज्यामयी । पराङेत्यनाशकी । इह चामुत्र चान्वेति । विद्वान्देवासुरानुभयान् ॥
1.27.12	यत्कुमारी मन्द्रयते । यद्योषिद्यत्पतिव्रता । अरिष्टं यत्किञ्च क्रियते । अग्निस्तदनुवेधति ॥
1.27.13	अश्रुतासश्श्रुतासश्च । यज्वानो येऽप्ययज्वनः । स्वर्यन्तो नापेक्षन्ते । इन्द्रमग्निश्च ये विदुः ॥
1.27.14	सिकता इव संयन्ति । रश्मिभिस्समुदीरिताः । अस्माल्लोकादमुष्माच्च । ऋषिभिरदात्पृश्निभि ॥
1.27.15	अपेत वीत वि च सर्पतातः । येऽत्र स्थ पुराणा ये च नूतनाः । अहोभिरद्भिरक्तुभिर्व्यक्तम् । यमो ददात्ववसानमस्मै ॥
1.27.16	नृ मुणन्तु नृ पात्वर्यः । अकृष्टा ये च कृष्टजाः । कुमारीषु कनीनीषु । जारिणीषु च ये हिताः ॥
1.27.17	रेतःपीता आण्डपीताः । अङ्गारेषु च ये हुताः । उभयान्पुत्रपौत्रकान् । युवेऽहं यमराजगान् ॥
1.27.18	शतमिन्नु शरदः ॥
1.27.19	अदो यद्ब्रह्म विलबम् । पितृणाञ्च यमस्य च । वरुणस्याश्विनोरग्नेः । मरुताञ्च विहायसाम् ॥
1.27.20	कामप्रयवणं मे अस्तु । स ह्येवास्मि सनातनः । इति नाको ब्रह्मिश्रवो रायो धनम् । पुत्रानापो देवीरिहाहिता ॥
1.28.1	विशीर्ष्णी गृध्रशीर्ष्णीश्च । अपेतो निर्ऋतिꣳ हथः । परिबाधꣳ श्वेतकुक्षम् । निजङ्घꣳ शबलोदरम् ॥
1.28.2	स तान्वाच्यायया सह । अग्ने नाशय संदृशः । ईर्ष्यासूये बुभुक्षाम् । मन्युं कृत्याञ्च दीधिरे । रथेन किꣳशुकावता । अग्ने नाशय संदृशः ॥

Reference	**Text *Devanāgarī***
1.29.1	प॒र्जन्या॑य॒ प्रगा॑यत । दि॒वस्पु॒त्राय॑ मी॒ढषे᳚ । स नो॑ य॒वस॑मि॒च्छतु ॥
1.29.2	इ॒दं वचः॑ प॒र्जन्या॑य स्व॒राजे᳚ । हृ॒दो अ॒स्त्वन्त॑र॒न्तद्यु॑योत । म॒योभू॒र्वातो॑ वि॒श्वकृ॑ष्टयस्सन्त्व॒स्मे । सु॒पि॒प्प॒ला ओष॑धीर्दे॒वगो॑पाः ॥
1.29.3	यो गर्भ॒मोष॑धीनाम् । गवा᳚ङ्कृ॒णोत्यर्व॑ताम् । प॒र्जन्यः॑ पुरु॒षीणा᳚म् ॥
1.30.1	पुन॑र्मामैत्विन्द्रि॒यम् । पुन॒रायुः॒ पुन॒र्भगः॑ । पुन॒र्ब्राह्म॑णमैतु मा । पुन॒र्द्रवि॑णमैतु॒ मा ॥
1.30.2	यन्मे॒ऽद्य रेतः॑ पृथि॒वीमस्का᳚न् । यदोष॑धीर॒प्यस॑र॒द्यदापः॑ । इ॒दन्तत्पुन॒राद॑दे । दी॒र्घा॒युत्त्वाय॒ वर्च॑से ॥
1.30.3	यन्मे॒ रेतः॑ प्रसि॑च्यते । यन्म॒ आजा॑यते॒ पुनः॑ । तेन॑ माम॒मृत॑ङ्कुरु । तेन॑ सुप्र॒जस॑ङ्कुरु ॥
1.31.1	अ॒द्भ्यस्ति॒रोधा॒ऽजा॑यत । तव॑ वैश्रव॒णस्स॑दा । ति॒रोधे॑हि स॒पत्ना᳚न्नः । ये अ॒पोऽश्न॑न्ति केच॒न ॥
1.31.2	त्वा॒ष्ट्रीं मा॒यां वै᳚श्रव॒णः । रथ॑ꣳसहस्र॒वन्धुरम् । पु॒रु॒श्च॒क्रꣳस॒हस्रा᳚श्वम् । आस्था॒याया॑हि नो ब॒लिम् ॥
1.31.3	यस्मै᳚ भू॒तानि॑ ब॒लिमाव॑हन्ति । धन॒ङ्गवो॑ ह॒स्ति॒हिर॑ण्यमश्वा॑न् । असा॑म सु॒मतौ॑ य॒ग्यिय॑स्य । श्रिय॒ बिभ्र॒तोऽन्न॑मुखीं वि॒राज᳚म् ॥
1.31.4	सु॒द॒र्श॒ने च॒ क्रौ॒ञ्चे च॑ । मै॒ना॒गे च॑ म॒हागि॑रौ । स॒त॒द्वा॒ट्टा॒रग॑मन्ता । स॒ꣳहा॒र्यन्न॒गर॒न्तव॑ ॥
1.31.5	इति मन्त्राः᳚ । कल्पो॑ऽत ऊ॒र्ध्वम् ॥
1.31.6	यदि॒ बलि॑ꣳहरे᳚त् । हि॒रण्यना॑भये वितु॒दये॑ कौबे॒राया॒यं ब॒लिः । स॒र्वभू॒तधि॑पतये॒ नम॒ इ॒ति । अथ ब॒लि॒ꣳहृ॒त्वोप॑तिष्ठे॒त ॥

Reference	**Text *Devanāgarī***
1.31.7	क्षत्रं क्षत्रं वैश्रवणः । ब्राह्मणां वयग्ग्स्मः । नमंस्ते अस्तु मा मां हिꣳसीः । अस्मात्प्रविस्यान्नंमद्धीति ॥
1.31.8	अथ तमग्निमांदधीत । यस्मिन्नेतत्कर्म प्रंयुञ्जीत ॥
1.31.9	तिरोधा भूः । तिरोधा भुवंः । तिरोधास्स्वंः । तिरोधा भूर्भुवस्स्वंः । सर्वेषां लोकानामाधिपत्यें सीदेति ॥
1.31.10	अथ तमग्निंमिन्धीत । यस्मिन्नेतत्कर्म प्रंयुञ्जीत ॥
1.31.11	तिरोधा भूस्स्वाहा । तिरोधा भुवस्स्वाहा । तिरोधा स्वंस्स्वाहा । तिरोधा भूर्भुवस्स्वस्स्वाहा ॥
1.31.12	यस्मिन्नस्य काले सर्वा आहुतीर्हुता भवेयुः । अपि ब्राह्मणंमुखीनाः । तस्मिन्नह्नः काले प्रंयुञ्जीत । परंस्सुप्तजंनाद्वेपि ॥
1.31.13	मास्म प्रमाद्यन्तंमाध्यापयेत् । सर्वार्थाः सिद्ध्यन्ते । य एवं वेद । क्षुध्यन्निदंमजानताम् । सर्वार्था नं सिद्ध्यन्ते ॥
1.31.14	यस्ते विद्यातुंको भ्राता । ममान्तर्हृदये श्रितः । तस्मां इममग्रपिण्डञ्जुहोमि । सर्मेऽर्थान्मा विवंधीत् । मयि स्वाहा ॥
1.31.15	राजाधिराजायं प्रसह्यसाहिने । नमों वयं वैश्रवणायं कुर्महे । स मे कामान्कामकामांय मह्यम् । कामेश्वरो वैश्रवणो दंदातु । कुबेरायं वैश्रवणायं । महाराजायं नमः । केतवो अरुंणासश्च । ऋषयो वातंरशनाः । प्रतिष्ठाꣳ शतधां हि । समाहितासो सहस्रधायंसम् । शिवा नश्शन्तंमा भवन्तु । दिव्या आप ओषंधयः । सुमृडीका सरंस्वति । मा ते व्योंम संदृशि ॥
1.32.1	संवत्सरमेतंद्व्रतञ्चरेत् । द्वौ वा मासौ ॥

Reference	**Text *Devanāgarī***
1.32.2	तस्मिन्नियमंविशेषाः । त्रिषवणमुदकोंपस्पर्शी । चतुर्थकालपानभक्तस्स्यात् । अहरहर्वा भैंक्षमश्नीयात् । औदम्बरीभिः समिद्भिरग्निं परिचरेत् । पुनर्मा मैत्विन्द्रिय मित्येतेनाऽनुंवाकेन । उद्धतपरिपूताभिरद्भिः कार्यङ्कुर्वीत । अंसञ्चयवान् । अग्नये वायवे सूर्याय । ब्रह्मणे प्रंजापतये । चन्द्रमसे नंक्षत्रेभ्यः । ऋतुभ्यस्संवथ्सराय । वरुणायारुणायेति व्रंतहोमाः । प्रवर्ग्यवदादेशः । अरुणाः काँण्डऋषयः ॥
1.32.3	अरण्येंऽधीयीरन् । भद्रं कर्णेभिरिति द्वे जपित्वा । महानम्नीभिरुदक॰ संग्ग्स्पर्श्य । तमाचाँर्यो दद्यात् । शिवा नश्शन्तमेत्योषधींरालभते । सुमृडीकेति भूमिम् । एवमंपवर्गे । धेंनुर्दक्षिणा । क॰सं वासंश्च क्षौमम् । अन्यंद्वाशुक्लम् । यंथाशक्ति वा । एवग्ग्स्वाद्ध्यायंधर्मेण । अरण्येंऽधीयीत । तपस्वी पुण्यो भवति तपस्वी पुंण्यो भवति ॥

Bibliography

Introduction

1. Alladi, Mahadeva Sastri, and K., Ranga'cha'rya. The Taittiríyáranyaka with the Commentary of Bhattabhaskara Misra Vol. I, Sanskrit. Mysore: Government Branch Press, 1900.
2. Apte, Hari Narayana. Taittariyaranyakam Bhashya of Sayanacharya, Sanskrit. Pune: Anandashram Sanskrit Granthavali, 1898.
3. Caland, W. Baudhayana Srauta Sutra, Vol. I, Sanskrit. Calcutta: Asiatic Society, 1904.
4. Caland, W. Baudhayana Srauta Sutra, Vol. II, Sanskrit. Calcutta: Asiatic Society, 1904.
5. Caland, W. Baudhayana Srauta Sutra, Vol. III, Sanskrit. Calcutta: Asiatic Society, 1904.
6. Challa, Lakshmi Nrusimha Sastry. Aruna Mantrartha Prakashika, Telugu. Machilipatnam: Aryananda Mudraksharashala, 1935.
7. Garbe, Richard. Srauta Sutra of Apastamba with a commentary of Rudradutta Vol. I, Sanskrit. Calcutta: Asiatic Society, 1882.
8. Garbe, Richard. Srauta Sutra of Apastamba with a commentary of Rudradutta Vol. II, Sanskrit. Calcutta: Asiatic Society, 1882.
9. Jamadagni, Shrikant, and Kashyap, R. L. Taittariyaranyaka Part 1, English. Bangalore: Sri Aurobindo Kapali Sastry Institute of Vedic Culture, 2014.
10. Kane, K. P., and Kane, P. V. Yajna, A Comprehensive Survey, English. Munger: Yoga Publication Trust, 2006.
11. Kashikar, C. G. The Baudhayana Srauta Sutra, Vol. I, English. New Delhi: Indira Gandhi National Centre for the Arts, 2003.
12. Kashikar, C. G. The Baudhayana Srauta Sutra, Vol. II, English. New Delhi: Indira Gandhi National Centre for the Arts, 2003.
13. Kashikar, C. G. The Baudhayana Srauta Sutra, Vol. III, English. New Delhi: Indira Gandhi National Centre for the Arts, 2003.
14. Kashikar, C. G. The Baudhayana Srauta Sutra, Vol. IV, English. New Delhi: Indira Gandhi National Centre for the Arts, 2003.
15. M. S., Ashwathanarayan Avadhani . Apastambiya Srauta Prayoga Vol. I, Sanskrit. Mattur: SrutiShankar Samskrit Samshodhana Pratishthanam, 2009.
16. M. S., Ashwathanarayan Avadhani . Apastambiya Srauta Prayoga Vol. II, Sanskrit. Mattur: SrutiShankar Samskrit Samshodhana Pratishthanam, 2009.

17. M. S., Ashwathanarayan Avadhani . Apastambiya Srauta Prayoga Vol. III, Sanskrit. Mattur: SrutiShankar Samskrit Samshodhana Pratishthanam, 2009.

18. Matysamahapuran, Hindi. Gorakhpur: Geeta Press, 2006.

19. Paaturi, Sitaramanjaneyulu. Matsya Mahapuranam, Telugu. Hyderabad: Venkateshwara Arshabharati Trust, n.d.

20. Pathak, Jamuna. Apastambasrautasutram, Dwitiyo Bhaga Hindi. Varanasi: Chaukhamba Sanskrit Series, 2014.

21. Pathak, Jamuna. Apastambasrautasutram, Prathamo Bhaga Hindi. Varanasi: Chaukhamba Sanskrit Series, 2014.

22. Pathak, Jamuna. Taittariyaranyakam Prathomo Bhaga, Hindi. Varanasi: Chaukhamba Sanskrit Series, 2014.

23. Ranade, H. G. Katyayana Srauta Sutra, English. Pune: Dr. H. G. Ranade and R. H. Ranade, 1978.

24. Shastri, Brahmalin Pandita Peetambaradutta. Srautayagnaprakriya Padarthanukramkoshah Dwitiyo Bhagah, Sanskrit. New Delhi: Rashtriya Sanskrit Sansthan, 2005.

25. Shastri, Brahmalin Pandita Peetambaradutta. Srautayagnaprakriya Padarthanukramkoshah Prathamo Bhagah, Sanskrit. New Delhi: Rashtriya Sanskrit Sansthan, 2005.

26. V., Sadagopan. Aruna Prasnam v1, English. Online: Sadogopan.org, n.d.

27. V., Sadagopan. Aruna Prasnam v2, English. Online: Sadogopan.org, n.d.

28. V., Sadagopan. Aruna Prasnam v3, English. Online: Sadogopan.org, n.d.

Ḇūmī

29. Abiogenesis, English. Online: Wikipedia, https://en.wikipedia.org/wiki/Abiogenesis, 2021.

30. Ashtekar, Abhay, Pawlowski, Tomasz, and Singh, Parampreet. Quantum Nature of the Big Bang, English. College Park: American Physical Society, 2006.

31. Big Bang, English. Online: NASA,https://science.nasa.gov/astrophysics/focus-areas/whatpowered-the-big-bang/.

32. Big Bang, English. Online: Wikipedia, https://en.wikipedia.org/wiki/Big_Bang, 2021.

33. Big Bounce, English. Online: Wikipedia, https://en.wikipedia.org/wiki/Big_Bounce, 2021.

34. Cyanobacteria, English. Online: Wikipedia, https://en.wikipedia.org/wiki/Cyanobacteria, 2021.

35. Ekpyrotic Universe, English. Online: Wikipedia, https://en.wikipedia.org/wiki/Ekpyrotic_universe, 2021.

36. Hydrothermal Vent Theory, English. Online: Wikipedia, https://en.wikipedia.org/wiki/ Hydrothermal_vent, 2021.

37. Lambda cold dark matter parametrization, English. Online: Wikipedia, https://en.wikipedia.org/wiki/Lambda-CDM_model, 2021.

38. Nebular hypothesis, English. Online: Wikipedia, https://en.wikipedia.org/wiki/Nebular_hypothesis, 2021.

39. Nucleogenesis, English. Online: Wikipedia, https://en.wikipedia.org/wiki/Nucleosynthesis, 2021.

40. Panspermia Theory, English. Online: Wikipedia, https://en.wikipedia.org/wiki/Panspermia, 2021.

41. Primordial Soup Theory, English. Online: Wikipedia, https://en.wikipedia.org/wiki/Primordial_soup, 2021.

42. Protoplanetary Disk, English. Online: Wikipedia, https://en.wikipedia.org/wiki/Protoplanetary_disk, 2021.

43. Theory of Evolution, English. Online: Wikipedia, https://en.wikipedia.org/wiki/Evolution, 2021.

Vāyumaṇḍalam

44. Aeronomy, English. Online: Wikipedia, https://en.wikipedia.org/wiki/Aeronomy, 2021.

45. Atmosphere of earth, English. Online: Wikipedia, https://en.wikipedia.org/wiki/Atmosphere_of_Earth, 2021.

46. Climatology, English. Online: Wikipedia, https://en.wikipedia.org/wiki/Climatology, 2021.

47. Jeannie, Allen. Tango in the Atmosphere: Ozone and Climate Change, English. Online: NASA, https://www.giss.nasa.gov/research/features/200402_tango/, 2004.

Varṣā

48. Cellular Metabolism, English. Online: Wikipedia, https://en.wikipedia.org/wiki/Metabolism, 2021.

49. Cellular Respiration, English. Online: Wikipedia, https://en.wikipedia.org/wiki/Cellular_respiration, 2021.

50. Photosynthesis, English. Online: Wikipedia, https://en.wikipedia.org/wiki/Photosynthesis, 2021.

51. Quantum Fields Theory, English. Online: Wikipedia, https://en.wikipedia.org/wiki/Quan- tum_field_theory, 2021.

52. Quantum Particle , English. Online: Wikipedia, https://en.wikipedia.org/wiki/Self-energy, 2021.

53. Water Cycle, English. Online: Wikipedia, https://en.wikipedia.org/wiki/Water_cycle, 2021.

Gurutvākarṣaṇa

54. Fundamental Forces, English. Online: Wikipedia, https://en.wikipedia.org/wiki/Fundamental_interaction, 2021.

55. General Theory of Relativity (General Relativity), English. Online: Wikipedia, https:// en.wikipedia.org/wiki/General_relativity, 2021.

56. Gravitational Waves, English. Online: Wikipedia, https://en.wikipedia.org/wiki/Gravitational_wave, 2021.

57. Standard model, English. Online: Wikipedia, https://en.wikipedia.org/wiki/Standard_ Model, 2021.

58. Quantum Gravity, English. Online: Wikipedia, https://en.wikipedia.org/wiki/Quantum_ gravity, 2021.

Sūrya Raṣmi

59. Electromagnetic Spectrum, English. Online: Wikipedia, https://en.wikipedia.org/wiki/ Electromagnetic_spectrum, 2021.

60. Photosynthesis, English. Online: Wikipedia, https://en.wikipedia.org/wiki/Photosynthesis, 2021.

61. Polar lights (Aurora), English. Online: Wikipedia, https://en.wikipedia.org/wiki/Aurora, 2021.

62. Sun's Radiation, English. Online: Wikipedia, https://en.wikipedia.org/wiki/Sunlight, 2021.

Yajṇenabanḍu

63. Bell, John S. On the Einstein Podolsky Rosen paradox. New Jersey: Physics Physique

Fizika, 1964.

64. Changanti, Venkata, Cheruvu, Murali, and Munnagala, Shastry V. Vedic Pravargya, a Thermodynamic Process, That Boosts Immunity, Reduces Pollution, and Mitigate COVID-19 Like Viruses, English. Online: International Journal of Scientific and Research Publications, 2021.

65. Einstein, Albert, Podolsky, Boris, and Rosen, Nathan. Can Quantum-Mechanical Description of Physical Reality be Considered Complete?. Minneapolis: Physical Review, v47 n10, 1935.

66. Euclidian space, English. Online: Wikipedia, https://en.wikipedia.org/wiki/Euclidean_ space, 2021.

67. Fadel, Matteo, and Et al. Spatial entanglement patterns and Einstein-Podolsky-Rosen steering in Bose-Einstein condensates. Washington D.C.: Science, 2018.

68. Gauger, Erik M, and Et al. Sustained Quantum Coherence and Entanglement in the Avian Compass. Online: Physical Review Letters, v106 n4, 2011..

69. Hensen, Bas, Hanson, Ronald, Et al. Loophole-free Bell-inequality violation using electron spins separated by 1.3 kilometres. London: Nature, 2015.

70. Higgins, Jacob S, and Et al. Photosynthesis tunes quantum-mechanical mixing of electronic and vibrational states to steer exciton energy transfer. Washington D.C.: Proceedings of the National Academy of Sciences of the United States of America, v118 n11, 2021.

71. Kolesnikov, Alexander I, and Et al. Quantum Tunneling of Water in Beryl: A New State of the Water Molecule. Online: Physical Review Letters, v116 n16, 2016.

72. Kong, Et al. Measurement-induced, spatially-extended entanglement in a hot, strongly-interacting atomic system., English. Online: Nature Communications, https://doi.org/10.1038/s41467-020-15899-1, 2020.

73. Kunkel, Philipp, and Et al. Spatially distributed multipartite entanglement enables EPR steering of atomic clouds. Washington D.C.: Science, 2018.

74. Lange, Karsten, and Et al. Entanglement between two spatially separated atomic modes. Washington D.C.: Science, 2018.

75. Marletto, Chiara, and Et al. Entanglement between living bacteria and quantized light witnessed by Rabi splitting. Bristol: Journal of Physics Communications, 2018.

76. Objective-Collapse Theory, English. Online: Wikipedia, https://en.wikipedia.org/wiki/ Objective-collapse_theory, 2021.

77. Quantum Entanglement, English. Online: Wikipedia, https://en.wikipedia.org/wiki/Quantum_entanglement, 2021.

78. Quantum Physics, English. Online: Wikipedia, https://en.wikipedia.org/wiki/Quan-

tum_ mechanics, 2021.

79. Reiter, George F, and Et al. Quantum Coherence and Temperature Dependence of the Anomalous State of Nanoconfined Water in Carbon Nanotubes.: The Journal of Physical Chemistry Letters, v7 n22, 2016.

80. Sarovor, Mohan, and Et al. Quantum entanglement in photosynthetic light-harvesting complexes. Berkeley: eScholarship, University of California, 2009.

81. Susskind, Leonard, Quantum Entangelments: Part 1, Online, Stanford University, https://www.youtube.com/playlist?list=PLA27CEA1B8B27EB67, 2006.

Index

संस्कृतम्

अग्नि, 27

अग्नेयाः, 65

अग्न्याधान, 126

अच्छावाक, 103

अणु, 83

अनादि, 6

अनुपलब्धि, 6

अनुमान, 6

अपौरुषेय, 6

अरणि, 126

अरुणकेतुका, 27

अरुणकेतुकाग्नि, 5

अरुणप्रश्न, 5

अरुणा, 25

अर्थशास्त्र, 63

अर्थापत्ति, 6

अवास्थ्यः, 140

अश्मिविद्विशः, 64

अश्वत्थ, 126

अश्विनौ, 63

असुरा, 28

अहङ्कार, 101

आकाश, 33

आग्नीध्र, 103

आत्म, 106

आदित्य, 26

आप, 33

आयुर्वेद, 3, 99

आरण्यकम्, 4

आरोग, 99

आह्वानीय, 139

इतिहासौ, 6

इन्द्र, 19, 27, 104

इन्द्रियाः, 101

उत्तरवेदि, 139

उपनिषद, 4

उपमान, 6

ऋग्वेद, 19

ऋतम्, 4

ऋतुसंहार, 63

ऋषि, 4

कर्मकाण्ड, 3

कल्पसूत्राः, 4, 125

काम्य, 125

कालिदास, 63

कृष्णयजुर्वेद, 5

केतवा, 25

क्श्यप, 100

गर्ग, 102

गार्हपत्य, 139

गुरुत्वाकर्षण, 7

गृहमेध, 64

गृहस्थ, 55

गृह्यकर्म, 125

गृह्यसूत्राः, 125

घनपाठी, 128

घर्म, 55

चयन, 138

चाणक्य, 63

चातुरमास्य, 63

ज्योतिष्मान्, 99

तन्मात्राः, 101

तपस्या, 21

तैत्तिरीय आरण्यकम्, 5

त्रेताग्नि, 139

त्वष्ट, 19

दक्षिणाग्नि, 139

दर्शपूर्णमासिष्टि, 125

दृष्टान्त, 7

देवता, 26

द्यौष्पितृ, 52

द्रव्यस्वभाव, 65, 75

द्विवचन, 83

धूपयः, 64

धृव, 99

नामधेय, 64

नित्य, 125

नित्याग्निहोत्र, 125

निमेष, 82

नियमाः, 141

निरूढपशुबन्ध, 63

नेष्ट, 103

नैमित्तिक, 125

पक्ष, 82

पञ्चकर्ण, 101

पञ्चभूताः, 23

पतंग, 99

पतर, 99

पद्धतयः, 4

परमाणु, 83

परमात्मा, 28

पर्जन्यः, 63

पश्यक, 93

पितराः, 27

पिशाचा, 28

पुराणाः, 6

पूर्वमीमांसा, 6

पूशन्, 27

पृथ्वी, 33

पोता, 103

प्रजापति, 19

प्रतिज्ञा, 7

प्रत्यक्ष, 6

प्रपाठक, 5

प्रमाणाः, 6

प्रयोगाः, 4
प्रवर्ग्य, 48
प्राचीनावंश, 139
प्राणत्रात, 102
प्रशास्ता, 103
प्लाक्षि, 101
बुद्धि, 101
बृहस्पति, 48
बौधायन श्रौतसूत्र, 139
ब्रह्मण, 19
ब्रह्मणाच्छन्सि, 103
ब्राह्मणम्, 4
भारतम्, 3
भूमि, 7, 104
भ्राज, 99
मत्स्य पुराण, 90
मनस, 101
मन्त्र, 23
मन्त्रपुश्पम्, 138
मयूखाः, 86
मरुताः, 63
महत, 101
महाभारता, 99
महामेरु, 100
महावेदि, 139
मित्रः, 63
मीमांसकाः, 6

मुहूर्त, 82

मूर्ति, 143

यजमान, 100

यज्ञ, 5

यज्ञेनबन्धु, 7

योग, 3, 99

रथचक्रं, 139

रसाः, 36

राक्षसा, 28

रुद्र, 19

लोकाः, 64

वज्र, 104

वत्स, 84

वराहव, 64

वरुण, 57

वरुणप्रघास, 63

वर्षा, 7

वागम्भृणी, 19

वातः, 90

वातराशना, 25

वायु, 27

वायु गणाः, 64

वायुमण्डलम्, 7

वालखिल्याः, 25

वास्तुशास्त्र, 3, 99

विध्युन्महसः, 64

विभास, 99

विराट, 26

विश्वेदेव, 47

विष्णु, 19

विहार, 138

वेद, 3, 21, 22

वेदाङ्ग, 4

वैखानसा, 25

वैशंपायन, 104

व्रत, 63

शब्द, 6

शमि, 126

शम्बर उदक, 69

शम्यु, 48

शाखा, 23

श्रुति, 6

श्रौत, 5

श्वापयः, 64

षोडशकर्मानि, 125

संध्यावंदन, 143

संसकृतम्, 63

संहिता, 4

सत्य, 7

सदस, 139

सप्त सूर्याः, 99

सप्तकर्ण, 101

सभ्यः, 140

सरस्वती, 63

सलिलम्, 21
सवितर, 85
सायणाचार्यभाष्य, 5
सावित्र चयन, 138
सूक्त, 19
सूर्य देवत, 99
सूर्यरश्मि, 7
सोम, 100
सोमयाग, 125
स्तुति, 63
स्मृति, 6
स्वतपसः, 64
स्वर्णर, 99
हविर्धान, 139
हेतु, 7
होता, 103

Devanāgarī Transliterated

Acçāvāka, 103

Āditya, 26, 39

Agneyās, 65, 75

Agni, 27, 33, 39, 63, 90, 106, 125

Āgnīḍra, 103

Agnyāḍāna, 126

Ahaṇkāra, 101

Āhvānīya, 139

Ākāṣa, 33

Anādi, 6

Aṇu, 83

Anumāna, 6, 7

Anupalabḍi, 6

Anuvākā, 23, 24, 25, 26, 27, 28, 29, 33, 36, 47, 48, 63, 64, 65, 66, 82, 83, 84, 85, 88, 93, 99, 100, 101, 102, 103, 104, 105, 109, 138, 139

Āpa:, 33

Apóruṣeya, 6

Araṇi, 126

Āraṇyakam, 4, 52, 55

Āroga, 99, 100, 102, 109, 113

Arṯāpatti, 6

Arṯaṣāśtra, 63

Aruṇā, 25

Aruṇaketuka, 138, 139

Aruṇaketukāgni, 5, 19, 47, 48, 52, 53, 85, 90, 99, 103, 104, 123, 138, 139, 141, 144

Aruṇapraṣna, 5, 6, 7, 19, 20, 23, 31, 33, 35, 36, 37, 39, 40, 41, 47, 49, 52, 55, 57, 58, 63, 67, 69, 70, 72, 75, 76, 82, 86, 88, 90, 93, 94, 99, 103, 107, 109, 111, 113, 114, 123, 138,

139, 141, 144

Aṣmivid̍viṣa, 64, 72

Asurās, 28, 39

Aṣvat̍t̍ạ, 126

Aṣvins, 63, 65, 66, 70

Āt̍ma, 106

Avāsṭ̍ya, 140

Āyurved̍a, 3, 63, 99

Ḅārat̍am, 3

Ḅāṣya, 5, 6, 35, 52, 90, 138, 139

Bo̍d̍āyana Ṣro̍t̍asūt̍ra, 139

Brahmaṇa, 19

Brahmaṇāçansi, 103

Brāhmaṇam, 4

Ḅrāja, 99, 102, 109, 113

Bṛhaspat̍i, 48, 52

Bud̍d̍i, 101

Ḅūmi, 52, 93

Cāṇakya, 63

Cāt̍uramāsya, 63

Cayana, 138, 139

D̍akṣiṇāgni, 139

D̍arṣapūrṇamāsiṣṭ̍i, 125

D̍evas, 27

D̍evat̍ā, 26, 27, 35, 39, 40, 52, 57, 63, 88, 99, 100, 103, 114, 125

D̍ravyasvaḅāva, 65

D̍ṛṣṭ̍ānt̍a, 7

Ḍṛva, 99

Ḍūpaya, 64, 72

Ḋvivacana, 83

Ḋyoṣpitṛ, 52

Ganapātīs, 128

Gaṇas, 64, 67, 72, 76

Gandarvās, 57

Garga, 102

Gārhapatya, 139

Garma, 55

Gṛhameda, 64, 72

Gṛhasṭa, 55

Gṛhyakarma, 125

Gṛhyasūtras, 125

Gurutvākarṣaṇa, 7, 77, 82, 86

Havirdāna, 139

Hetu, 7

Hotā, 103

Indra, 19, 27, 39, 47, 48, 52, 57, 63, 104, 109

Indriyas, 101

Itihāsas, 6

Jyotiṣmān, 99, 102, 109, 113

Kālidāsa, 63

Kalpasūtras, 4, 125

Kāmya, 125, 138

Karmakāṇda, 3

Ketavā, 25

Kṛṣṇayajurveda, 5

Kṣyapa, 100, 101, 102, 111, 114

Lokās, 64

Mahābāratā, 99

Mahāmeru, 100, 101, 102, 103, 111, 113, 114

Mahata, 101

Mahāvedi, 139

Manasa, 101

Maṇdalas, 105, 106

Mantra, 23, 24, 25, 26, 27, 28, 29, 33, 47, 48, 63, 64, 65, 66, 70, 82, 83, 84, 85, 99, 100, 101, 102, 103, 104, 105, 138

Mantrapuṣpam, 138

Marutas, 52, 63

Matsya Purāṇa, 90

Mayūkas, 86, 90

Mīmā˙sakas, 6, 7

Mitra, 63, 106

Muhūrta, 82

Mūrti, 143

Nāmadeya, 64

Nemittika, 125

Neṣṭa, 103

Nimeṣa, 82

Nirūdapaṣubanda, 63

Nitya, 125

Nityāgnihotra, 125

Niyamas, 141

Paddatis, 4

Paksa, 82

Pancabūtā, 23

Pancabūtās, 33

Pancakarna, 101, 102

Paramānu, 82

Paramātmā, 28

Parjanyā, 63

Pasyaka, 93

Pata˙ga, 99, 102, 113

Patara, 99, 102, 109, 113

Pisācas, 28

Pitaras, 27, 39, 40

Plāksi, 101

Potā, 103

Prācīnāva˙sa, 139

Prajāpati, 19, 23, 24, 25, 26, 29, 35, 36, 37, 39, 40, 41, 47

Pramānas, 6

Prānatrāta, 102

Prapātaka, 5, 23, 24, 25, 26, 27, 28, 29, 47, 48, 63, 64, 65, 66, 82, 83, 84, 85, 99, 100, 101, 102, 103, 104, 105

Pratijnā, 7

Pratyaksa, 6

Pravargya, 48, 49, 52, 55, 58

Prayogas, 4

Prsāstā, 103

Prtvī, 33

Purāṇas, 6, 41, 99

Pūrvamīmā˙sā, 6, 19

Pūṣan, 27, 39, 40

Rākṣasās, 39

Rasā, 36

Raṭacakra˙, 139

Ṛgveda, 19, 20, 35, 128

Ṛṣi, 84, 101, 102, 104

Ṛtam, 4, 29, 55

Ṛtusa˙hāra, 63

Rudra, 19, 47, 48, 49, 51, 52, 53, 55, 57, 58

Sa˙ḍyāva˙dana, 143

Sa˙hitā, 4

Sa˙sakṛtam, 63

Ṣabda, 6

Saḇya, 140

Sadasa, 139

Ṣāḵā, 23, 28, 29

Salilam, 21, 23, 31, 33, 35, 36, 41

Ṣambara Udaka, 69

Ṣami, 126

Ṣamyu, 48

Saptasūryas, 99, 100, 101, 102, 103, 104, 107, 109, 111, 113, 114

Saptakarṇa, 101, 102

Sarasvatī, 63

Satya, 7

Savitara, 85, 90

Sāvitra Cayana, 138, 139

Sāyaṇācārya, 5, 6, 52, 72, 82, 90, 93, 138, 139

Smṛti, 6

Ṣodaṣakarma, 125

Soma, 100, 103, 111, 114

Somayāga, 52, 125

Ṣrota, 63

Ṣruti, 6

Stuti, 63

Sūkta, 19, 20, 35

Sūrya Devatā, 99

Sūryaraṣmi, 7, 95, 99, 107

Ṣvāpaya, 64, 72

Svarṇara, 99, 102, 109, 113

Svatapasa, 64, 72

Tanmātras, 101

Tapasyā, 21, 25, 35

Tettirīya Āraṇyakam, 5

Tretāgni, 139, 140

Tvaṣṭṛ, 19

Udaka, 69

Upamāna, 6

Upaniṣad, 4

Uttaravedi, 139

Vāgambṛṇī, 19

Vajra, 104

Vālakilyā, 25

Varāhava, 64

Varṣā, 7, 59, 63, 67, 69, 72, 75

Varuṇa, 57, 63, 106

Varuṇapragāsa, 63

Vāstuṣāstra, 3, 99

Vāta, 90

Vātarāṣanā, 25

Vatsa, 84, 101

Vāyu, 27, 33, 39, 63, 64, 67, 72, 76, 85, 90

Vāyu Gaṇas, 64, 72

Vāyumaṇdalam, 7, 40, 43, 47, 49, 51, 55, 57

Veda, 3, 4, 5, 6, 7, 19, 41, 52, 55, 58, 63, 99, 109, 123, 144

Vedāṇga, 4

Vekānasā, 25

Veṣa`pāyana, 104

Vibāsa, 99, 102, 109, 113

Vidyunmahasa, 64, 72

Vihāra, 138, 139

Virāta, 26

Viṣṇu, 19, 52, 53, 84, 85, 86, 88, 90, 94

Viṣvedeva, 47

Vratas, 63

Yajamāna, 100, 103

Yajṇa, 5, 7, 48, 63, 69, 99, 103, 123, 125, 126, 127, 128, 138, 141, 142, 143, 144

Yajṇapuruṣa, 52

Yajṇenabandu, 7, 115

Yoga, 3, 99, 123, 143

English

Abiogenesis, 15, 17

Agnostic, 19

Amino acids, 15

Anthropomorphic, 35

Apathetic, 19

Atmosphere, 15, 40, 45, 46, 51, 52, 53, 55, 57, 58, 60, 62, 69, 70, 72, 75, 76, 78, 98, 113, 114, 141

Atmospheric layers, 5, 7, 47, 49, 57, 75, 98, 107, 113

Axial tilt, 78

Axis of rotation, 39

Baryogenesis, 36, 39, 40

Beryllium, 13

Big bang theory, 11

Big bounce, 11

Biosphere, 60, 72

Carbon, 13

Catalyze, 52

Cellular life, 15, 141

Cellular metabolism, 60

Cellular respiration, 45, 60

Chlorofluorocarbons, 52

Classical newtonian physics, 117

Cloud micro physics, 73

Cloud physics, 72

Cognitive sciences, 5, 35

Collapse models, 117

Condensation, 6, 46, 60, 62, 67, 72

Conductive transfer, 46

Consciousness, 19, 29, 35, 36, 37

Core, 15, 33, 125, 140

Cosmic radiation, 45

Cosmic ray fission, 13

Cosmogony, 93, 138

Cosmology, 7, 19, 33, 36, 39, 93, 138

Crust, 15

Cryptography, 122

Cyanobacteria, 15, 45

Dark energy, 81

Dark matter, 11, 81

Detection efficiency, 120

Deuterium, 13

Deuterium oxide, 36

Diurnal, 4

Diurnal temperature variation, 45

Dogmatic, 19

East-west winds, 45

Ekpyrotic, 11

Electromagnetic force, 78, 117

Electromagnetic radiation, 97, 107, 111, 114, 118

Electron cloud, 120

Electronic transition frequency, 80

Electrostatic forces, 13

Elementary particle, 118

Empirical, 58, 76

Energy field, 72

Epistemology, 6

Euclidian space, 118

Evaporate, 60, 83

Evaporation, 60, 69, 75

Exegesis, 6

Falsifiability, 5

Field, 72, 81, 86, 94, 111, 114, 122

Frequency, 80, 96, 97

Gaussian curves, 119

General theory of relativity, 81

Gluons, 13

Gravitational collapse, 11

Gravitational forces, 13

Graviton, 78

Gravity, 5, 7, 13, 62, 81, 93, 117

Gravity waves, 45

Greenhouse effect, 45

Halons, 52

Heavier elements, 13

Helium, 46, 120

Hertz, 97

Human anatomy, 128

Human physiology, 128

Hydrogen, 13, 36, 46, 120

Hydrosphere, 60

Hydrothermal vent theory, 17

Hypothesis, 7, 15, 69, 70, 78, 109, 122, 126, 142

Inorganic matter, 15

Inter-specie evolution, 15

Interstellar gas and dust, 11, 33

Iron, 45

Isotopes, 13

Lambda cold dark matter parametrization, 11

Latitude, 45, 51

Laws of motion, 6

Lighter elements, 13, 36

Lithium, 13

Lithosphere, 60, 62

Local hidden variables, 120

Locality, 120

Longitude, 51

Loop quantum gravity, 33

Macrocosm, 19, 126, 127, 138, 141, 142

Magnetosphere, 46

Mantle, 15, 36

Metaphysics, 35

Meteorites, 15

Microcosm, 19, 138

Microwaves, 97

Molecular mass, 46

Natural sciences, 13

Nebular hypothesis, 11, 39

Noctilucent clouds, 46

Nuclei, 13, 46

Nucleogenesis, 13

Organic, 15

Organism, 15, 17

Orthopraxical, 5

Oxygen, 15, 36, 45, 47, 51, 52, 53, 58, 62, 121

Ozone, 15, 45, 47, 49, 51, 52, 53, 55, 58

Ozone disassociation, 53

Ozone-oxygen interconversion cycle, 50, 53

Panspermia theory, 17

Parametrization, 11

Particle, 72, 78, 118, 119

Particle physics, 118

Photoionization, 46

Photoelectric effect, 6

Photon, 118

Photosynthesis, 40, 45, 60, 70, 97, 120, 122, 128

Planetary waves, 45

Planetesimals, 15

Postulate, 11, 35, 52, 69, 138, 139, 143

Potentiality, 20, 21

Precipitation, 60, 62, 67, 69

Primordial, 17

Protoplanetary disk, 11, 13

Quantum entanglement, 69, 118, 119, 120, 121, 122, 123, 126, 127, 138, 139, 141, 143, 144

Quantum physics, 5, 7, 11, 35, 119

Quantum state, 118

Quantum, 33, 36, 67, 69, 70, 72, 76, 81, 86, 90, 94, 117, 118, 119, 120, 121, 122, 123, 126, 127, 128, 138, 139, 140, 141, 143, 144

Quantum bridge, 33

Quantum computing, 122

Quantum cryptography, 122

Quantum decoherence, 121

Quantum fields, 67, 72, 76

Quantum geometry, 33

Quantum gravity, 90

Quantum teleportation, 122

Quantum tunneling, 140

Quarks, 13

Radiometric dating, 13

Repeatability, 5

Revolution, 78, 97, 117

Runoff, 62

Singularity, 11

Solar electromagnetic radiation, 5, 7

Solar nebula, 11

Solar system, 11, 13, 15, 93, 111

Spacetime fabric, 78, 80

Standard model, 78

Strong nuclear force, 78, 117

Sublimate, 62

Supernovae, 13

Synodic rotation, 97

Theistic, 19

Theory of evolution, 15, 17

Thermonuclear fusion, 13

Transpiration, 60, 62, 69

Tritium, 13

Velocity, 80

Water cycle, 45, 60, 62, 72, 93

Weak nuclear force, 78, 117

About Author

Sitaram Sarma Ayyagari is one of those guys who was lucky to be educated both in *Veda* and Science. Like many of his generation who had limited career paths, he had to satisfy his passion for core sciences while studying engineering. Bachelors in Electronics and Communications Engineering was followed by Masters in Electronics Instrumentation and finally culminated with an MBA to climb up the ladder in Corporate America. The informal and probably more important education he got was in *Veda*. He underwent initial training in portions of *Kṛṣṇayajurveda* and *Ṛgveda* from his father, Sri Sreerama Murty, and his uncle Sri Bheemasankaram. Later, he was trained with Pandit Sri Ramesh Rajamani in Dayton, Ohio, and from Late *Brahmaṣrī Ganāpāṭī* Gullapalli Udayakumar Sarma in Houston. Meanwhile, his insatiable scientific curiosity fueled his self-study all along, and it was in quantum physics that he discovered a connection to the *Vedic* Rituals. Sitaram is the co-founder of Bharateeya Vaahini, a non-profit organization that promotes the learning of *Veda Vāṇmaya* with a scientific perspective.

Kovidanam Vani
www.kovidanamvani.com

www.ingramcontent.com/pod-product-compliance
Lightning Source LLC
Chambersburg PA
CBHW021836220426
43663CB00005B/276
9780981759609